T0092416

The Possibility of Life

THE POSSIBILITY OF LIFE

SCIENCE, IMAGINATION, AND OUR QUEST FOR KINSHIP IN THE COSMOS

JAIME GREEN

THORNDIKE PRESS
A part of Gale, a Cengage Company

Portions of this work appeared, sometimes in different form, in "If the Earth Isn't Special, Then the Whole Cosmos Is," *Slate* (2021); "The Strangely Human Messages We Send To Aliens," Medium (2018); "How We Imagine Aliens," *Medium* (2018); "Could astrobiology research convince us to fight climate change?" *Popular Science* online (2018).

Thorndike Press, a part of Gale, a Cengage Company.

Thorndike Press® Large Print Nonfiction.

The text of this Large Print edition is unabridged.

Other aspects of the book may vary from the original edition.

LIBRARY OF CONGRESS CIP DATA ON FILE.
CATALOGUING IN PUBLICATION FOR THIS BOOK
IS AVAILABLE FROM THE LIBRARY OF CONGRESS.

ISBN-13: 979-8-88579-465-7 (hardcover alk. paper)

Set in 16 pt. Plantin.

Published in 2023 by arrangement with Harlequin Enterprises ULC.

Printed in Mexico

Printed Number: 1 Print Year: 2024

ADDITIONAL COPYRIGHT INFORMATION

In memory of my Zaide, Norman Epner.
And for Miles Nova, everything new.

In memory of my Zaide, Norman Epner.
And for Miles Nova: everything new.

Living, being in the world, was a much greater and stranger thing than she had ever dreamed.

–Ursula K. Le Guin, *The Tombs of Atuan*

TABLE OF CONTENTS

WATCHFUL STARS

When I was very little, I was scared of the night sky. I remember hurrying from my parents' car to the front door so that I wouldn't be out in the open too long. I felt like the stars were watching me.

But soon enough, the sky became a friendlier place. My dad taught me the names of the constellations and how to find the North Star pouring off the tip of the Big Dipper (a real feat in the light-polluted skies of Queens). And then, a few years later, the idea of a vast cosmos became a promise, not a threat, because it might be full of benevolent creatures. What happened was I started watching *Star Trek: The Next Generation*.

Those are the first aliens I remember — well, the first I remember loving, because I was terrified of ET — but quickly my world became full of them. Kind Starfleet officers with minor prosthetics to differentiate them from their human colleagues; the sandworms of Arrakis; glowing deep-sea angels in *The Abyss;* Meg Murry's beloved Aunt Beast. There were bad guys and monsters, but there was so much hope, too. And while pop culture was populating my imaginary worlds, the science I learned in school and from PBS specials watched on playdates with equally nerdy friends was expanding my sense of what was possible. Rovers landed on Mars, SETI listened for signals, and, around when I was twelve years old, the first planets beyond our solar system were found.

Thinking about aliens was in some ways the same thing as thinking about science. Looking at a star and imagining planets around it was the same as imagining who might live there. Learning about space-time and the hard limit of light speed was the same as imagining ways to subvert these laws to traverse the galaxy, was the same as thinking about my friends on the *Enterprise* who already had. Thinking about aliens was thinking about whether life really needed water and carbon, or eyes and hands. Thinking about aliens who might be plants or bugs was thinking about the possible inner life of the plants and bugs in my backyard.

But it was more, of course. Thinking about who might be out there was thinking about the possibilities of existence, and how humans fit into all of it.

Some people are drawn to science by their drive to understand, but what I have always loved most is how science shows us what we don't know, how little we understand of the world even as we're inextricably a part of it. I didn't want to answer questions but discover mysteries instead — mysteries with tantalizing possibilities, theories and hypotheses and whispers of the truth.

Science reveals that our world is so much more than what we see every day. There are cells churning beneath our skin, bizarre ecosystems hidden at the bottom of the ocean, and clockwork mysteries ticking away in an atom's heart. But space offers possibly infinite real estate for wonder. In 1995, the Hubble Space Telescope aimed its mirrored eye at a dark spot in the constellation Ursa Major; it found, in this empty patch of sky, nearly three thousand galaxies. This photograph, the Hubble Deep Field, has always captivated me: clouds and spirals scattered in silver, amber, and red — an impossible richness you could blot out if you held your thumb up to the sky.

We're now at the brink of being able to answer questions that have obsessed humans since we've known how to ask. Does life exist beyond what we know? Is life in the cosmos common, rare, or even unique? Are we alone?

Except we've felt that we're on the brink of these answers for decades. As science makes concrete progress — telling us more about the planets around other stars, delving into the workings of the human brain — the big answers seem held on the tip of the universe's tongue. Scientists who scan the sky for alien signals seem to always expect an answer within the next ten to twenty-five years, whether they're postulating from 1980 or 2023. Most every Mars rover launches with the same hope, that this time we'll finally find evidence of life on our neighbor planet, which in its past may have been habitable, but today seems starkly not.

And so, in the absence of answers, scientists fall back on the next best thing: odds. They untangle paradoxes with equations and approximations. They take what we know, multiply it by an estimate, and home in on a spot on the spectrum — common or special, likely or rare.

To be sure, odds are useful in science. Before you can go looking for something, or get the funding to do so, it's helpful to be able to show there's a decent chance that it's out there. And there are mathematically rigorous ways to approach it.

The most famous way of thinking about the odds of alien life is the Drake Equation. Except, in practice, it hardly works as an equation at all, and it provides no definite answers. Social scientist and NASA consultant Linda Billings more accurately calls it the "Drake Heuristic," an approximation that reveals as much about the analyst's assumptions as it does about the question being asked.

That question is this: How many civilizations are out there in the galaxy, right now, with technology that we could detect from Earth?

Frank Drake, a pioneer in the field of SETI (the Search for ExtraTerrestrial Intelligence), didn't even devise his equation to try to answer that question. Instead, it was an agenda-setting exercise for an early meeting of SETI practitioners, outlining the discussion in terms of which variables determine who might be out there possibly sending a signal.

The variables:

R_* The rate at which stars form.

f_p The fraction of stars that have planets.

n_e The average number of planets per star that could support life.

f_l The fraction of those planets that develop life.

f_i The fraction of planets with life where life develops intelligence.

f_c The fraction of intelligent civilizations that develop detectable technology.

L The average life span of a signaling civilization — how long they produce signals that we could detect.

All of those multiply to give you N, the number of civilizations in the galaxy whose transmissions we could currently, conceivably, find.

$$N = R_* \cdot f_p \cdot n_e \cdot f_l \cdot f_i \cdot f_c \cdot L$$

In the decades since Drake wrote that out on a chalkboard, the equation has become a ubiquitous and contested tool. It is attacked for its blind spots, the many gaps between, say, the origin of life and the emergence of intelligence. Or for its generalizations — what is intelligence, anyway? And astronomers deploy it to illustrate the rationality of their pessimism or optimism: *Plug in these very reasonable numbers, and see how the output plummets or soars!*

But what Drake was really saying was that in order to think about SETI, about the search for signs of alien intelligence, you need to think about all these things: the formation and evolution of stars and planets, the origin of life, the emergence of intelligence, the nature of technology, and the longevity of civilizations. There isn't an answer at the end of his equation, just a host of new questions to ask.

But the question most often — too often — asked about extraterrestrial life is *whether or not*. The fact is, that's a very boring question. Its *answer* would be revolutionary, and plenty of scientists are pursuing it, but sitting here posing it to a vast and empty sky doesn't tell us anything.

We should really be asking *what if?*

What if? is not a linear path to knowledge but a circular one, spiraling recursively toward deeper and deeper understanding. When we imagine extraterrestrial life, we strain against the limits of scientific knowledge, going out on a limb and testing its strength, then looking back at ourselves from a new distance. *What if?* makes us ask questions about life on Earth. How did it arise? Why is it the way it is? Are we special and what would that mean? What is our responsibility with the intelligence and technology that we have?

What if? is not an unscientific question. It drives every hypothesis and prediction, every leap and act of synthesis that moves us from the unknown toward knowing. Scientists imagine things every day. They imagine a possible chemical pathway and test it. They imagine the surface of a distant planet about which hardly anything is known and render it as artwork. They imagine the evolutionary lineage of a parallel Earth, the near future, the outcome of unprecedented events.

We imagine alien life in a similar way through fiction. You know this. You've read the novels and seen the movies. Sometimes these stories seem fantastical (blue-skinned cat-people who live among floating mountains), sometimes they seem like an eerily plausible vision of future events (we receive a signal from a distant star, and everyone here fights about what to do with it). And all these works are keys to understanding the cosmos just as much as scientific inquiry.

There's more to writing aliens in fiction

than deciding who says *Take me to your leader.* There are questions of biology and astronomy, as well as sociology, linguistics, philosophy — questions that transcend the realms of science. Questions like *Does the past reflect possible futures? How much does our experience constrain imagination?* And *What do we want humanity's future to be?* In this way science fiction is more than entertainment, it's a generative act that creates new possibilities of life beyond Earth, as valid and potent as anything we might conjure up in the lab. Through fiction we can move beyond likelihoods and binary outcomes to look instead at what our imaginations do with the limitless possibilities of outer space and, crucially, ask what that might mean.

Space is indescribably vast. Planets orbit, stars burn, and the black hole at the center of the Milky Way devours, all without a care for our existence. We're just a speck, just a flicker of matter organizing itself in a funny way for a bit of time.

We take our smallness as proof of both our insignificance and our great importance. For all that we imagine, Earth is the only planet we know of with life. It's a terrible burden to contemplate, that this might be the only life in the cosmos. A huge responsibility for such a clumsy species to bear, and such a lonely universe to live in.

We want to know if we matter. We want to know what our existence means. Life on another world would give us a context, a foil, a richer way of understanding ourselves. So we search and listen, calculate the odds. But ultimately we can only speculate about how we fit into the galactic picture.

Through it all, our visions of alien worlds are reflections of ourselves, arising from our research, our dreams, and our subconscious like mist from a field at dawn. When we imagine a dozen ways for an alien to be, we're imagining a dozen different kinds of people. When we invent alien languages, we learn more about the human brain. When we dream of a benevolent visitation, we're telling a story of what we think we need.

In a novel about imagination shaping the world, *The Lathe of Heaven,* Ursula K. Le Guin wrote, "What will the creature made all of seadrift do on the dry sand of daylight; what will the mind do, each morning, waking?" If these imaginings are dreams, the science and stories both, what stays with us when we wake?

Imagining extraterrestrial life is a way of figuring out what it means to be a conscious animal, what it means to be matter and alive. Our visions of space are a reflection of our selves and of our humanity — like the building blocks of a telescope, a mirror and a lens.

CHAPTER 1

ORIGINS

Of course, we start with *Star Trek*.

Specifically in the white rock canyon of a fossilized seabed of a distant planet, in an episode of *Star Trek: The Next Generation* called "The Chase." An away team from the *Enterprise* is in a standoff with a pair of Cardassians and four Romulans, alien species traditionally hostile to our heroes.

The *Enterprise*'s mission, in case you don't have the opening credits narration committed to memory, is "to seek out new life and new civilizations, to boldly go where no one has gone before." It's not a warship or a science vessel but some hybrid or transcendence of those categories. The crew of the *Enterprise* seeks knowledge and offers protection to those in need, occasionally gets involved in skirmishes but never seeks them out.

But today, tensions run high. The three parties have all come to this site on the hunt for a precious fragment: ancient DNA that holds the missing piece of a molecular puzzle that was scattered, four billion years ago, across a quarter of the galaxy. A message is embedded in the genetic code of nineteen different species. The *Enterprise* crew has been chasing these fragments, hoping to complete the puzzle and decode the message.

"This is not a natural design," Commander La Forge told Captain Picard earlier in the episode. "This is part of an algorithm, coded at the molecular level."

Picard says, "So, four billion years ago someone scattered this genetic material into the primordial soup of at least nineteen different planets across the galaxy?"

Lieutenant Commander Data, the android crew member, offers, "The genetic information must have been incorporated into the earliest life-forms on these planets and then passed down through each generation."

The ship's doctor, Beverly Crusher, asks why anyone would want to do this. Captain Picard adds, "And what was this program designed to do?"

La Forge says, "Well, we couldn't know that until we assembled the entire program and then ran it."

And so, they do, collecting bits and pieces

of genetic code from the diverse species that crew the starship and determining that the last missing piece is from this planet where they now stand. But the Romulans and Cardassians have been on the hunt, too, leading to this stalemate.

The *Enterprise* away team is two humans and two Klingons; the humans look like the humans you know, the Klingons' high hairlines reveal corrugated foreheads. The Cardassians have cartilaginous ridges running up the sides of their necks, around their eyes, and reaching up their foreheads. The Romulans' ears are pointed, and their eyebrows sweep up diagonally. Everyone, to your ears, speaks English, thanks to universal translators, aside from a Klingon's occasional expletive or insult (*toppa'!*).

The *Enterprise* crew seeks information, but the Romulans and Cardassians seek status, too: ownership of the prized knowledge and control of whatever the algorithm reveals. So while they try to intimidate each other, Dr. Crusher unobtrusively scrapes a bit of the ancient seabed into a vial and lays it across the captain's tricorder. He whispers, "The program has been activated. I think it's reconfiguring the tricorder . . . to project something."

Out of the tricorder comes an image of a

humanoid being, the same size and shape as the humans, Klingons, Cardassians, and Romulans assembled there. (The same size and shape as so many of the other species peopling the series and films of the Star Trek world, because it's the size and shape of a human actor.) This being has no forehead ridges . . . no brow ridge at all, even at a human scale. She has a smooth, mottled head and tiny cauliflower ears, and she wears a plain white tunic — altogether the epitome of a neutral base from which other species can be extrapolated. She says:

> Life evolved on my planet before all others in this part of the galaxy. We left our world, explored the stars, and found none like ourselves. Our civilization thrived for ages, but what is the life of one race, compared to the vast stretches of cosmic time? We knew that one day we would be gone, that nothing of us would survive. So, we left you.

This first sentient species seeded the galaxy, not with life itself but little tweaks to various nascent genomes so that all those worlds would give rise to people who looked like them. Heads, arms, faces, bodies, all in echo of their gardener-ancestors. And alongside those instructions, fragments of this message. The ancient ones hoped that their scattered children would, through the requisite cooperation, come together in harmony to hear their message.

It's a bit like finding out God exists. Or maybe it's even more meaningful because your creator isn't an all-powerful being but just someone like you, someone who came before and made you in their image, yes, but so that the universe might continue to be known and explored, so that life's legacy might continue. And they not only made you but made the many species you've come to know across the galaxy. Not God's children but kindred, nonetheless.

Yet this revelation falls on unreceptive ears. The blustery Klingon captain who's been traveling with the *Enterprise* scoffs, "That's all? If she were not dead, I would kill her." The Cardassian sniffs, "The very notion that a Cardassian could have anything in common with a Klingon . . . it turns my stomach." And Captain Picard sort of shruggingly requests his team to be beamed back aboard.

Back in his quarters, Picard sits fiddling with the ceramic artifact that kicked the episode off, brought to him by his old archaeology professor, the man who'd been pursuing the mystery of these fragments. It's a toaster-sized sculpture of a humanoid, the top half of which comes off to reveal a clutch of smaller figures within. As Picard tells his second-in-command, "The Kurlan civilization believed that an individual was a community of individuals. Inside us are many voices, each with its own desires, its own style, its own view of the world."

The ancient alien hologram had told its assembled seekers, "There is something of us in each of you, and so, something of you in each other." The ceramic figure represents a cosmic community, a galactic family, and the smaller figures within it are each a species, each a world.

Perhaps this episode is just an elaborate retcon to explain why there are so many humanoid aliens in the *Star Trek* universe. But it also reflects a whole constellation of hopes and fears and questions we have, here on Earth today. The hologram ancestor says, "We left our world, explored the stars, and found none like ourselves." We fear that, as we learn more about what's beyond our solar system, silence and loneliness await us. She also says, "Our civilization thrived for ages," a luminous hope to a viewer today, in the face of climate change and pandemics and nuclear weapons. And the product of her people's work, a galaxy brimming with intelligent life, life that can find each other and recognize each other and, with the help of a little technology, meet and speak — what more could we hope for when we look at the stars?

But how are we to get there? To a populated cosmos, to that kinship and communication? How did we get *here,* to our inhabited planet, smart enough at least to worry we're the only ones?

Before we can explore these questions, let's look to their roots. For storytelling, this means our culture's earliest imaginings of life on other worlds and the explosions of these kinds of stories that followed pivotal discoveries about the solar system. And for science, it means trying to understand how life arises — how it began, in the past, on Earth, and how that informs our understanding of what we seek on other planets. Searching for life elsewhere forces us to confront our limited understanding of what life even is.

A Brief History of Life on Other Worlds

The idea that we might be alone in the galaxy is a relatively new one. In the last century and a half, the more we've learned about the universe, the harder it has become to imagine it full of life. It's only in the last thirty years that the pendulum has begun to swing back, as telescopes reveal thousands of planets around other stars and the building-block molecules of life abundant in the nebulae of interstellar space.

But during the European Renaissance, new discoveries cracked open the cosmos, filling it with possibility. When Copernicus realized the Earth wasn't the center of the universe, when Galileo discovered that the untwinkling wandering stars were *planets* and that the Milky Way, a smear of light in the sky, was actually full of stars, the universe seemed fat with potential. But then we learned the details — the airless worlds, the baking atmospheres, the stars flinging sterilizing rays at all their planets. As our telescopes got better, and our rovers got on their rockets, we learned the lonelier truth.

Centuries since Copernicus, we've learned that not only is the solar system *not* the center of the galaxy but our galaxy isn't the center of the universe and our universe may not even be all there is. We're still guided by what's called the Copernican principle, the idea that in no realm should we take humanity to be special.

Scientists often refer to the *n=1 problem,*[1] the challenge of extrapolating from one example (ours) to anything about life elsewhere. If we only know how life is *here,* we have nothing against which to triangulate. But for centuries, n=1 was a gift of sorts. We knew only one world — Earth — so that became the model for what other worlds would be like. Specifically: inhabited.

Visions of inhabited worlds were first arrived at philosophically, derived more from metaphysics than science. Aristotle, most influentially, placed the stationary Earth at the center of a set of nested spheres. It was an easy assumption from his point of view, matching daily human experience if not the details of astronomical observation. But what was most important was that this arrangement reflected his sense of humanity's place — central — and his understanding of matter and physics.

This geocentric worldview, later bolstered by Ptolemy, would be adopted by the Christian church through the Middle Ages and a bit beyond, supporting quite nicely the idea of humanity as God's focus and primary creation. But it has never been the only way to

1 Where *n* stands for the number of samples in the pool.

see the world. Other Greek thinkers bought into the idea of plurality. The Atomists, in the fourth and fifth centuries BCE, who believed that all matter was made of indivisible atoms (for matter could not be infinitely divisible into smaller and smaller parts) thought the universe contained more atoms than could be used up building our one *kosmos;* thus, there must be infinite worlds. One Atomist, Metrodorus of Chios, even leaped from the idea of an infinite universe to an infinity of species. "To consider the Earth as the only populated world in infinite space is as absurd as to assert that in an entire field sown with millet, only one grain will grow." Pythagoreans speculated that the moon was inhabited, its native animals larger and more beautiful than ours. Buddhists understand the universe to be composed of six inhabited realms,[2] all equally real, into any of which a soul can be reborn.

In the West, Aristotle's model carried the meaning philosophers and priests needed to be true. As a scientific model, though, it was always a little wonky. In the second century BCE, the Greek philosopher Aristarchus bucked consensus, asserting that the Earth orbited the sun, in part because he

2 Known by their inhabitants: gods, demigods, humans, animals, hungry spirits, and demons.

worked out that the sun was so much larger; he also deduced that the stars were incredibly distant, while Aristotle preached that they were no farther away than the visible planets like Venus and Mars. The geocentric model was also suspiciously complicated. Observed from Earth, planets sometimes seem to change direction and start moving backward in the sky, against the backdrop of stars. We now know this is due to the dance of our orbit and theirs, sometimes aligned and sometimes moving in opposition, but geocentrism had no such explanation. Instead, the model accounted for that retrograde motion with a complicated solution called epicycles, having planets not orbit the Earth directly but instead orbit a point that then orbited the Earth.

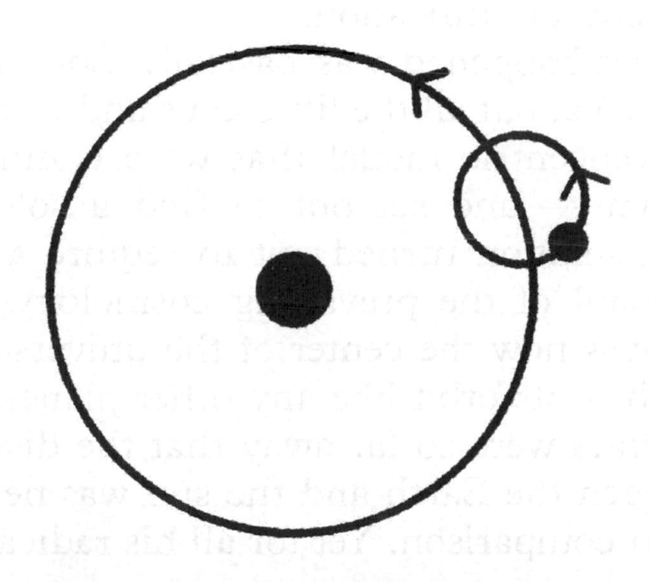

And even with those concessions, it was nearly impossible to use the geocentric model to make accurate predictions, a major red flag. As astronomer Caleb Scharf puts it in his book, *The Copernicus Complex*, "Planets would arrive a bit early or late to certain positions in the sky." And for fourteen hundred years, he writes, there was not even one universally accepted model of geocentrism. The big picture was consistent, but depending on whether you prioritized philosophical clarity or mathematical precision, you'd tweak things here and there to suit your needs. By medieval times, Scharf writes, "cosmological models were in a thoroughly messy and inconsistent state." Ironically, when Copernicus shattered the reigning worldview, he also brought us closer to a conceptual unification.

What happened was basically Copernicus learned about all the little gaps and errors in the geocentric model that were commonly known — and set out to find a solution. That solution turned out to require a total overhaul of the prevailing cosmology. The sun was now the center of the universe, the Earth in its orbit like any other planet, and the stars were so far away that the distance between the Earth and the sun was negligible in comparison. Yet for all his radicalism,

Copernicus did still hold on to some old-fashioned, incorrect ideas. He, like Aristotle, was beholden to the idea of circular orbits and constant velocities for their "perfection," so he had to still rely on epicycles to match the observed paths of the planets. Johannes Kepler, working half a century later, would be the one to do the math that replaced pesky circular orbits with their mathematically accurate elliptical routes.

About a century after Copernicus first laid out his theory,[3] Galileo built his telescopes, a radical advance on those that had come before, and through them he found visual evidence to support Copernicus's mathematical calculations: the planets were worlds, sometimes orbited by their own moons (quite different from our own), and the smudge of the Milky Way was, in fact, a density of stars too thick to be made sense of by the unaided eye. The skies were not a shadow show put on for humanity's benefit; they were a mess

3 Copernicus first described his heliocentric theory in a manuscript that only circulated privately in 1514; he published his landmark *De revolutionibus orbium coelestium (On the Revolutions of the Celestial Spheres)* in 1543; Galileo built his first telescope in 1609 and made his momentous discoveries in 1610.

of stars and planets, a universe impossibly vast, viewed from just another one of their number.

The specks in the sky suddenly expanded to 3D, no longer points but spheres. Historian of science Steven J. Dick writes, "Copernicus gave birth to a new tradition where the term *world* (*mundus*) was now redefined to be an Earthlike planet . . . If the Earth is a planet, then the planets may be Earths; if the Earth is not central, then neither is man." The syllogism continued: if Earth is a planet like all others, then all these other planets would be made of the same stuff. Earth, air, fire, water, for a baroque example. And the spark of life, too — there's no reason to think that mysterious substance, which was believed to be very much a substance, would be only here. Kepler, for example, believed that the matter of Earth possessed a "vital faculty," an intrinsic life force that gave rise to plants and animals. So, Kepler wrote, "whatever faculty this Earth, being one of those bodies, possesses, it is reasonable to suppose that the others, too, are provided with such faculties."

This historical question and its evolution are called the *plurality of worlds*. It wasn't a question of whether there are other planets or other solar systems — *worlds* here isn't *planets* but something more like *homes*, places where life could exist. The assumption was almost unstated, that where there was a niche, life would fill it.

It was only thanks to scientific discoveries that these worlds could be imagined. Before Copernicus and Galileo revealed the nature of the solar system, writes anthropologist John Traphagan, who studies the relationship between culture, science, and religion, "the capacity of Europeans to imagine other worlds with intelligent beings, or even other worlds beyond the observable realm of the planets, was very limited." It wasn't just that they didn't have the necessary facts. Instead, Traphagan writes, they were limited by the boundaries a culture places on what is imaginable. Our inventive abilities can really only extend so far beyond the real. Copernicus's genius for shattering these confines only followed after a series of mathematical solutions created a new possibility. And so it is in storytelling, too. People weren't imagining aliens on other planets because other planets weren't understood as such. But once other worlds were made available to imagination, those stories came in like a flood.

Before Copernicus, the sun and the moon had been occasional settings for satire or utopias or purely fanciful forays. But after

Copernicus, as scholar of culture Karl S. Guthke writes in his book, *The Last Frontier,* the new reality of other worlds makes these fictions each "a search for potential truth."

The truth these writers sought in the cosmos was that of a deeper understanding of life on Earth. The geocentric model hadn't been mathematically pretty or smooth, but it had been *meaningful.* The church had doubled down on Aristotelian thinking because it not only suited but manifested Christian doctrine. We understood humanity as central to God's concern and to the movement of the heavens. Copernicus upended far more than geometry.

Heresies break open doors, and sneaking in behind them come a flood of other new notions. And so in the wake of Copernicus and Galileo, the plurality of worlds flourished as both a scientific idea and a fictional inspiration. The line between science and fiction wasn't even concerning. For centuries, scientific communication was intertwined with fictional imaginings, even more than it is today, with scientists, writers, and those in the murky in-between moving from astronomy to philosophy to fabulist fiction from one treatise to the next and within a single work.

Carl Sagan was hardly sui generis as an

astronomer using fiction to explore the power of scientific discoveries. Fiction helps us integrate new information and make sense of things. *Okay, so* this *is what it feels like to believe the cosmos has other worlds,* a test for our sea legs when we find ourselves at sea.

The first world that post-Copernican writers populated was the moon. Kepler himself wrote of a journey to the moon in *Somnium* (*A Dream*), published in 1634, which Guthke identifies as "the first literary work that, inspired by Copernican science, treats of aliens in outer space." *Somnium* is a bit of a dumpling story, a book wrapped within a dream within a book: the narrator dreams that he is reading a book, and within that book a spirit from the moon appears to him.

The spirit speaks on space travel, lunar geography, and his fellow lunar inhabitants — flora, fauna, and human — and explains how life adapts to the moon's long days and Earth- and space-facing halves. It's couched in scientific fact (and mathematical rigor), but just as intensely thought out are the imaginings of lunar society. Those on the Earth-facing side mark time by the Earth's waxing and waning, and they live in a climate made temperate by the planet's reflected sunlight. Those living on the far side of the moon are nomadic, so as to cope with their hemisphere's more extreme conditions. And over the whole moon, Kepler imagines, life is larger than what we know on Earth.[4] He was just as inspired by Galileo as by Copernicus, writing in 1611 that "it is . . . as though Galileo had opened a new gate to the heavens, through which one could now see with one's own eyes what was previously hidden." Part of that revelation was the moon's landscape, seen with telescopic magnification — its mountains and valleys so strikingly like Earth's. Kepler was moved to wonder why this world would exist if not to be inhabited. "May not some creatures less noble than man be imagined such as might inhabit those tracts?"

About thirty years later, English bishop Francis Godwin's *The Man in the Moone: or, A Discourse of a Voyage Thither,* imagined an explorer carried by birds to the moon. Godwin's moon is, like Kepler's, a supersized version of Earth, inhabited by giant plants, animals, and people. Unlike Kepler's moon people, though, Godwin's are decidedly

4 The moon's large mountains suggested to Kepler that life there, if it existed, would be proportionately larger than life on Earth — bigger mountains, bigger animals, bigger men.

superior to Earthly humans, virtuous to almost utopian extremes, living "in such love, peace, and amitie, as it seemeth to bee another Paradise."

Soon enough, observations through better telescopes suggested that the moon was not the habitable mini-Earth these stories imagined. But these same observers began to speculate that other planets could harbor life. In work published in 1698, the astronomer Christiaan Huygens posited that human life existed on all planets. A century later, astronomer William Herschel speculated "that the Sun's dark spots were actually holes in a glowing hot atmosphere, beneath which, a cool surface supported large alien beings," Caleb Scharf writes in *Scientific American*. With similar extravagance, a few decades after that, amateur astronomer Thomas Dick suggested that about eight million beings lived in Saturn's rings.

It was a moment of fabricated abundance. As intellectual minds moved away from religion, they found a new fellowship in the cosmos. But these visions also reflected the science of their time. Life, in the 1700s, was considered an extension of matter. If there was no divine breath required for animation, then life was spontaneous and everywhere.

These worlds were simply new *wheres* for it to be.

Then, the nineteenth century saw a pair of scientific theories that, together, offered a new vocabulary to these stories. Darwin's theory of evolution established a framework for life with both historical depth and imaginative breadth. In concert with that, Pierre-Simon Laplace's nebular hypothesis of planet formation (which built on earlier ideas from Kant) suggested not only that planets should be abundant, a natural consequence of stars' formation, but that planets, too, had an evolutionary trajectory. According to Laplace's model, the planets farther out from a star coalesced before those closer in, shifting the worlds of the solar system from analogously similar to now logically diverse. (This timeline also provided a new paradigm for the century's fiction: the idea that planets farther out from the sun would have had more time to evolve led to the soon-to-be-born genre of frighteningly superior Martian invaders.)

Scharf writes that from the time of Copernicus to the turn of the twentieth century, "the question of life beyond the Earth seems to have been less of 'if' and more of 'what'." Yes, there were objections — the plurality of worlds creates a whole host of thorny problems for the belief that Earth's people are

God's primary concern. Some philosophers remained steadfast in their anthropocentrism, even after the heliocentric model was proved true. But others embraced our new place in a cosmic brotherhood, whether that evoked for them the awe of grand connectedness or of our smallness in a vast world. But as early as the eighteenth century, the *whats* and *hows* gained ground on the *ifs*, and anyone who read or thought on the subject lived their lives believing we were not alone.

As our knowledge of the cosmos expanded, fewer scientists explored outer space through fiction. But those who did went fanciful. The work of astronomer Camille Flammarion is emblematic of this moment. Influenccd by Darwin, his *Lumen,* published in 1887, features a disembodied soul recounting his series of reincarnations across many alien worlds, describing how on each world life adapted to the planet's particular circumstances. In his essay, "Science Fiction before the Genre," Brian Stableford writes, "No other nineteenth-century work is so thoroughly imbued with a sense of wonder at the universe revealed by astronomy and the Earth sciences."

It wasn't all celebration and camaraderie, though. The cosmos could also be terrifying. The sky had once been a close dome but now was an infinite blackness. We once lived under God's gaze; now, if he even still existed, his attention was spread across infinite worlds. Writing in 1714, English essayist Joseph Addison was as frightened by the night sky as I was as a child: "I could not but look upon myself with secret horror, as a being that was not worth the smallest regard of one who had so great a work under his care and superintendency. I was afraid of being overlooked amidst the immensity of nature, and lost among the infinite variety of creatures, which in all probability swarm through all these immeasurable regions of matter." H. G. Wells wrote of a similar feeling in an essay from 1894: "There is a fear of the night that is begotten of ignorance and superstition, a nightmare fear, the fear of the impossible; and there is another fear of the night — of the starlit night — that comes with knowledge, when we see in its true proportion this little life of ours . . ."

Flammarion, though, reaches toward alien life in an ecstatic embrace: ". . . shall we greet them? My brothers, let us all greet them: those are our sister humanities passing by!"

What would it be like to live with the confident knowledge, as many did during the Enlightenment, that aliens existed? I marvel at it the way I marvel about assured religious faith. I guess that shows my hand: both feel frankly speculative to me. Imagine, I sometimes think, what it would be like to face death confident of an afterlife, especially one of eternal reward. I'd likely feel the same sense of equanimity knowing that the cosmos is full of intelligent life, of creatures, of ecosystems, of novel chemical solutions to turning the universe's energy into something ordered and persistent. That knowledge would change what it means to be human.

So these stories, and all the many others that have followed them, are where we try out living in that knowledge for a spell. Science fiction guides us as we seek to expand the frontiers of knowledge — not outward in space but inward in self-knowledge. What does it mean to be human?

The birth of modern science fiction, then, comes with a pair of imagined Martian invasions that together encapsulate so many of humanity's fears and hopes, and lay the foundation for the genre. In Kurd Lasswitz's *Auf zwei Planeten* (*Two Planets*), published in 1897, Martian invaders end up as "enlightened guardian angels," contrasting H. G. Wells's probably-more-familiar-to-you tentacled monster invaders from the same year.

It was the birth of a new paradigm — humans no longer journeying to other worlds to chronicle the oddities, instead now subject to predation by a superior species. Wells's Martians are physically monstrous, and Lasswitz's look like humans, but both authors write in the long-standing tradition of imagining Martians as superior — given their planet's age (per Laplace's theory), they've had more time to evolve.

Both authors use this superiority to question paths of supposed progress humanity has set for itself. The invasion in *The War of the Worlds* — technologically advanced Martians nearly conquering the Earth — shows humanity's penchant for conquest and colonialism through a new lens.[5] Lasswitz's

5 Wells even writes in the book's first chapter, "[B]efore we judge of [the Martians] too harshly we must remember what ruthless and utter destruction our own species has wrought, not only upon animals, such as the vanished bison and dodo, but upon its own inferior races. The Tasmanians, in spite of their human likeness, were entirely swept out of existence in a war of extermination waged by European immigrants, in the space of fifty years. Are we

superior aliens swoop in to give humanity a crash course in "ethical instruction," initially resisted as conquest but eventually accepted. But even these benevolent aliens have, in evolution, lost some of their so-called humanity, prefiguring the human/Vulcan dichotomy of *Star Trek* and countless other stories that define human nature as emotional and instinctive against a cold and logical other.

These stories set the mold for a century of sci-fi that would follow, stories of invasion and rescue, of alien others manifesting society's fears and hopes for who we might become. And good thing that the momentum was there, because in the twentieth century, science itself offered less reason to believe in the possibility of plurality. Laplace's theoretical abundance was replaced by decades of prevailing wisdom that planets were in fact extremely rare. And while for centuries the existence of extraterrestrial life had rested on the assumption that where there was appropriate matter, life sprung into being, sparked by some "vital force," Louis Pasteur disproved the theory with a few flasks of sterilized broth in 1859, and in

---

such apostles of mercy as to complain if the Martians warred in the same spirit?"

the mid-twentieth century the origin of life was understood to be a momentous event of complicated and lucky chemistry. Understanding how it happened on Earth, then, would be crucial to believing it could possibly ever happen anywhere else.

Histories of Life on Earth

Before the origin of life was a mystery, it was matter of fact. Into the nineteenth century, the theory of spontaneous generation held sway. Some living things were known to require parents: human beings, dogs, other things we'd seen being born or sprouting from seeds. But sometimes, it was thought, life just happened. When meat sat out too long, maggots started to emerge. A rotting log sprouted mushrooms. Life from not-life. Aristotle credited pneuma, or "vital heat," as the generative force. This vital heat had a soullike power and permeated the air. Pneuma animated matter, with different substrates giving rise to different kinds of life. That same force that was thought to fertilize the worlds beyond Earth was active here as well, sprouting flies from trash and mice from hay. And the scientific advances of the Renaissance seemed at first to bolster the theory. Newton revealed matter as imbued with gravitational power. Why would life not be such an intrinsic force as well?

The same year that Darwin published *On the Origin of Species,* 1859, Louis Pasteur put spontaneous generation to bed. He was less trying to disprove it than prove the validity of germ theory (which holds that diseases are caused by microorganisms too small to be seen), but the two were intertwined. Pasteur showed that sterilized beef broth generated nothing spontaneously, in conditions where broth usually would. It wasn't something about the broth that gave rise to life, it was something invisible living in it, something that could be boiled off and killed.

But while Pasteur proved that life didn't habitually arise from nonliving matter, we know that it must have once. Somewhere on a world of rock and water and space dust, some chemicals came together in a way that made it keep happening, snowballing into a system that used energy, stored information, and could grow, evolve, replicate, and persist.

What happened? How did inert matter turn into life? That's the impossible question, unless you have a time machine or, as one astronomer wistfully wished to me, unless aliens happened to have been observing

Earth at the time and can offer to share the tape. Or if they'll take you back there to see for yourself.

In the last episode of *Star Trek: The Next Generation,* the omnipotent alien Q has been zapping Captain Picard through time, into his past, present, and future. But he takes one meaningful detour, much farther back.

Q: Welcome home.

PICARD: Home?

Q: Don't you recognize your old stomping grounds? This is Earth, France, about, oh, three and a half billion years ago, give or take an eon or two. Smells awful, doesn't it? [He sneers like someone's forgotten to change a litterbox.] All that sulfur and volcanic ash. I really must speak to the maid.

It's night on the newly formed planet. Or still forming — rocks peek out of a sea in the distance, but closer to the cave where Picard and Q stand, red-orange lava ambles by. But these tumultuous conditions might be just what life needs, to be jostled into existence. Q clambers down toward a tarry tide pool. "Right here, life is about to form on this planet for the very first time." He dips his hand into the pool, pulling it back out covered in gloopy mess. "Too bad you didn't bring your microscope. It's really quite fascinating. Oh, look, there they go. The amino acids are moving closer and closer and closer."

In that moment, though, nothing happens, thanks to the spatial anomaly (don't worry, that doesn't mean anything) that Picard must unravel to save humanity (don't worry, he does). The emergence of life, it seems, was a fragile, fortuitous event.

Or perhaps new life wasn't so serendipitous. In *Star Trek*'s vision, the Earth's surface hasn't even yet cooled, roiling and spouting hot sparks just yards away from the fateful pool. And in reality, too, while we don't know where or how chemistry crossed that ineffable line to aliveness, we do know that it happened remarkably quickly, just about as soon as conditions on the new planet allowed.

And what *Star Trek* shows is much like what I remember learning in high school, of fortuitous reactions in a primordial soup. In 1952, chemist Stanley Miller[6] recreated his era's best understanding of the ancient

6 Who you may not know was a graduate student at the time; Harold Urey, credited as coauthor of the work, was his adviser.

Earth's chemistry in a flask, a mix of simple compounds like water, ammonia, methane, and hydrogen, and shot it through with electric zaps simulating lightning. Soon enough, the broth turned murky, thickening with amino acids, the building blocks of proteins, which are themselves one of the building blocks of life. No one crowed *It's alive!* And it wasn't, of course, but it felt like a momentous step, and the recreation of life's origin seemed to be within reach.

The first life left no fossils behind.[7] Early biochemical reactions did leave some tantalizing clues in the geological record, like iron oxide bands that indicate the appearance of oxygen in the atmosphere and thus the first opportunity for rust, but the specifics of what happened and where and how weren't recorded. So from Miller through to researchers today, scientists have come up with ways they think life might have arisen, and then they try to . . . do it again themselves.

Could these building blocks plus some lightning have resulted in life?

7 Scientists use "metabolic fossils" to refer to biochemical clues in contemporary life from which we can extrapolate distant ancestors' biochemistry and genes, but those are metaphorical fossils, not remnants in rock.

I dunno. Let's try it.

Millennia and a whole planet condensed to weeks or months in a lab.

Today, we know the building blocks Miller sought to synthesize are actually already abundant: interstellar dust is filthy with organic molecules, and meteorites crash to Earth laden with amino acids. But a shopping cart full of ingredients doesn't make a cake. So some researchers today focus less on building blocks than processes, the reactions and pathways that use and produce the various molecules we know to be components of biochemistry. Because it is the processes, not the molecules — the actions, not the matter — that makes something alive.

Current thinking on the origin of life breaks into two main camps, with a handful of alternatives on the fringes and in between. The predominant view is called RNA World, imagining a phase in life's ancient history before our genetic code became double-helixed. RNA is sort of half a molecule of DNA[8] — a backbone of sugar and phosphate, with nucleobases protruding like the teeth of a comb. (In DNA, a pair of these

8 With some minor chemical variations

combs connect and twist together into one double-helixed molecule.)

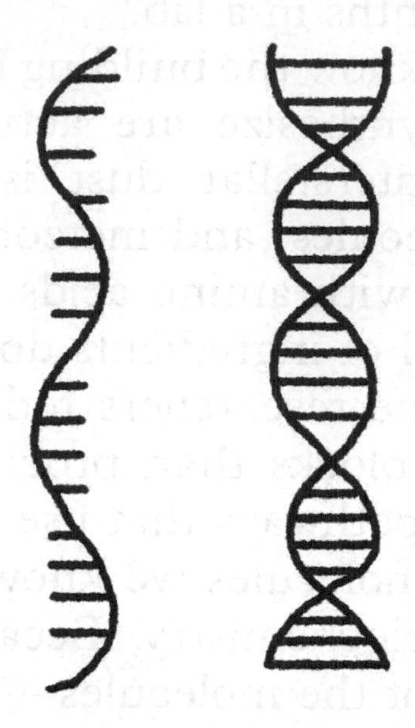

RNA is basically never in this untangled state, tending instead to fold and clump.

In our modern cells, RNA works mostly as a messenger, copying DNA's information and shuttling it to the cell's protein-making factories. But in the early 1960s, the geneticist J.B.S. Haldane got thinking about how some viruses don't use DNA for their genes but instead use RNA for everything. It's a simpler system, RNA as messenger and genome, and viruses seem so primitive — maybe RNA came first. Later it would be discovered that RNA molecules not only are messengers and code repositories but can act as enzymes, too, catalyzing chemical reactions. RNA seemed an appealing multipurpose tool, perhaps an origin before cellular chemistry became specialized.

Steven Benner was a pioneer in the field of synthetic biology, one of the first researchers to create an artificial gene. Now, his lab studies the origin of life — specifically, the origin of RNA. These early complex molecules are his holy grail. "If you can get to RNA on an abiological process," if you can make these molecules of life out of nonliving chemistry, "you're almost home."

He told me, "If I look at how you do things in your body today, and how plants do things, and how amoebas do things, you are all using RNA catalysts to make your protein, so RNA predates proteins by that model." In terms of the chemical evolution of life, RNA is as far back as scientists can see before things get swallowed up by the fog of the unknowable. And as an information carrier, it's one way of differentiating life's chemistry from everything else.

RNA World is a dominant-enough view, especially in the older generation of researchers, that when I spoke to scientists in that school they would phrase things like, *Well everyone agrees . . .* and *Everyone sees that the path is . . .* But the broader field is hardly unified. RNA World supposes that the first glimmers of life were molecules that carry information, but replicating information isn't the only thing life does. And the focus on RNA ignores the question of how that earliest life powered its replication and evolution, even its simple growth.

An alternate view came not from biology or chemistry but geology, from noticing how energy moves through rocks and how it moves through life, and seeing the possibility of parallels between the two.

In the 1980s, geologist Mike Russell began to suspect the existence of undersea hydrothermal vents that would form from the spewing of mineral-rich water into the cooler ocean. Unlike the high-energy deep-sea vents that had already been discovered, these alkaline vents would feature a gentler burbling. Minerals dissolved in the water would gradually precipitate into stone structures, like if stalagmites were chimneys and undersea.

Russell's hypothesized vents would solve a crucial question about the origins of life: Where does the energy come from to make all this new chemistry happen? Deep-sea alkaline vents would come with a built-in source: the moving water, for starters, and the difference between the alkaline flow and the more acidic seawater. (The difference in pH creates a gradient of potential chemical energy.) The stone structures have pores in their walls that would be containers to keep the churning matter concentrated in a small, cell-like space. (One problem with Miller and Urey's "hot, dilute soup" is that if it's the ocean, it's incredibly dilute, and if the building blocks of life aren't concentrated, there's little likelihood they'll interact.[9]) And in 2000, one of these ocean-floor vent systems was finally found to exist deep in the Atlantic. It was called the Lost City.

"Energy flux promotes self-organization of matter," biochemist Nick Lane writes in his book, *The Vital Question.* He points to organized states of matter that arise when energy flows through a system: hurricanes, ocean currents, the steady spiral of bathwater above an open drain. "This has nothing to do with information," Lane writes, but "it can create environments where the origin of biological information — replication and selection — is favored." All life on Earth, he points out, runs at the cellular level by the

9 For this reason, Benner and others think life may have had an easier time getting started on the drier — but not, in those eras, totally dry — Mars.

creation of energy gradients across a membrane. Differences in charge and chemical concentration between one side of a membrane and the other drive cellular machinery with a buildup of potential energy. That universality, to Lane, suggests that metabolism — the ways cells use energy — is at the root of the tree of life. (To be fair, also universal is the language of life's genetic code.)

Today, Lane is after the molecules that drive metabolism, working in the lab to prove his hypothesis by conjuring those molecules from an initial vent-style soup of carbon and hydrogen. Benner is, in parallel, pursuing RNA. They both start with chemical soups that seem plausible to the ancient Earth. Other researchers are wetting and drying lipid molecules that will pop into spherical membranes on their own, an origin path for the containers that allow life to be something distinct from its environment. Benner says, "One of the big puzzles in this business is how do you make long RNA molecules that have enough information that they might support Darwinism of some kind." That's his driving question. Lane says his inquiry is guided by biology. "What are the conditions that give rise to a self-replicating entity? What is the simplest self-replicating entity? . . . To my mind DNA is quite difficult. RNA is quite difficult. . . . How can something get better at making copies of itself before you have a code?" The thing being replicated doesn't even need to be a molecule but could even be a chemical-reaction cycle itself — if every time through the cycle, the component molecules are doubled, and on and on.

The rivalry between RNA World and metabolism-first researchers has calmed down in recent years, largely due to younger scientists in the field tending toward a more holistic approach. Chemist Kamila Muchowska, who counts herself in that category, told me that she and other researchers recognize "the simultaneous importance of genetics, metabolism, and potentially compartmentalization," which is the sequestering of bits of chemistry in isolated, concentrated pockets, because "there is no known life that relies on one of these and not on the other two." But even that approach is in some ways just another camp among many, with researchers attempting to find solid footing in a wide-open field.

William Martin, an evolutionary biologist and biochemist who collaborated with Russell on the theory of energy-driven origins, told me that the study of the origin of life is troubled by an almost total lack of constraints.

"There are almost no facts, okay? . . . All we have is logical inference." He says it's an essential human quality to want to understand our past, but that because these questions are so central to humanity, he's also noticed a trend in his field. When a young scientist has their first encounter with a theory about the origin of life, he said, "they assimilate it into the fabric of their thought structure." Anything after that, they take in light of the first theory they learned. And these ideas matter to people so deeply that it becomes very hard to change their minds.[10]

I found myself falling into that very trap, enamored with the elegance with which the first theory I encountered, alkaline vents, had been explained. It's a strange kind of science, I think, not trying to understand what *does* happen but what *did*. When I told Benner that it seemed like the origin of life was an unanswerable question, he said, "This question in the first place is confused. It's two questions. There's a historical question at some level: How did it happen? And that's actually a very challenging question to answer. Then there's the chemistry/geology question, which is, How might it have

10 "Sound familiar in the context of modern politics?" Martin added, and he was right.

happened?" These researchers aren't trying to prove how life started on Earth. They realize the futility of that question. Even with a perfectly complete chemical theory, replicated and proven in the lab, you could never confirm that it worked the same way on Earth those billions of years ago. The value is more in understanding the options and their implications. *Could* life arise like this? Not winnowing down the possibilities but manifesting them.

Synthetic biologist Kate Adamala's approach to the origin of life sidesteps all the historical questions that the metabolism-first and RNA World camps retread. "I think they're both right," she told me, "and they're both wrong. I think it was always a coevolution." It seems implausible to her that a proto-cell could get informationally or energetically complex without the other elements of life's functions we see today. Metabolism and information are the biggest, most obvious of life's gestures. But there are so many more, and we understand them so poorly.

We don't know everything that goes on in even the simplest cells — we don't know all the chemical components, let alone how they all function and interact. Adamala told me about the synthetic bacterium *Synthia,* built by researchers at the J. Craig Venter Institute, whose genome was assembled and sequenced piece by piece. And still, there were mysteries. "Even though they know every single gene in that system," she told me, "they still don't know what about sixty of those genes do."

Adamala wants to understand everything happening in the cells she works with so that she can start stripping them down. She's trying to reduce a cell to its simplest functioning form. All the other bells and whistles of complexity would then represent later evolutions. Instead of trying to restage a version of life's origin in the lab — "waiting for those wimpy ribozymes to finally figure out how to make proteins," as she put it — she comes at the origin from the other direction, taking a cell that already knows how to make proteins and stripping it down to bare bones. She hopes to be able to de-evolve it back all the way to "the root of the tree of life."

For now, she's working on the first half of that process, learning how to make a living system in the lab. ("You can actually touch the point when the dead goo becomes living goo, and that's the most fascinating to me.") Her synthetic cells have a membrane the same as living cells, with a simple protein channel that allows molecules to pass through. Inside is a patchworked genome, with instructions for the cell's activities, and about 400 other chemical compounds — proteins and the enzymes needed to make more of them — distilled and purified from natural sources. "Everything goes in one by one, and we know the concentrations of every single one of those components." Adamala said she likes the knowing, and the control — it's what her training as a chemist prepared her for.

That control lets Adamala know every chemical compound in her synthetic cells, and adding them one by one she can know their function. It's not just that living cells have more than 400 chemical components — they do: many, many times more — but we don't know the exact number because we don't even have a complete census.

At 400 components, Adamala's cells have enough chemistry to do a few things they're programmed for: exchange antibiotic resistance, fuse with other cells, or resist (similarly engineered) parasitism. Adamala calls them *toy systems*, each a test case for a function that a living cell would need, building different kinds of complexity into the cells so that eventually she can build a fully complex, fully functional cell — to strip it down to its barest basics.

But in that middle point, the living synthetic cell — how will Adamala know she's gotten there? "I know instinctively that what I'm working with is dead. I can look at *E. coli* or my dog and say, 'This is alive.'" She trusts that she'll know it when she sees it. "I'm not a philosopher, so I'm not good at discussing things. I'm good at pipetting." She hopes that at some point she'll pipette enough to know.

Sara Imari Walker, a physicist and sometime collaborator with Adamala, worries that *We'll know it when we see it* dooms our hopes to understand life, whether at its origins or on other worlds. *Life* is notoriously hard to define. NASA has a working definition: "Life is a self-sustaining chemical system capable of Darwinian evolution," but this excludes, as Walker has put it, "mules and senior citizens" and could thwart observers who don't have time to sit and watch for many generations as evolution takes place. For the 1970 edition of the *Encyclopaedia Britannica,* Carl Sagan wrote a survey of the possible kinds of definitions of life — physiological, metabolic, biochemical, genetic, and thermodynamic — and showed how there are exceptions to all, entities that meet the criteria but are clearly, to us, not alive, or vice versa. "An automobile, for example, can be said to eat, metabolize, excrete, breathe, move, and be responsive to external stimuli." It seems almost more alive than a dormant seed or spore. Except we know that it's not.

But the problem may not be finding the right definition. Philosopher Carol Cleland said that even attempting to define life shows that we're on the wrong track. She told *Astrobiology Magazine,* "Definitions tell us about the meanings of words in our language, as opposed to telling us about the nature of the world." She says that what we need is not a definition but a theory, to know not what the word *life* means to us, but what life *is*. But with neither a theory nor a definition, we're left with our intuition, entirely shaped by the life we know on Earth.

Walker worries that we are much further from understanding life in this fundamental way than we might think. She compares it to the state of human understanding of planetary movement centuries ago. Before Copernicus and Kepler, astronomers described planetary motion with epicycles, those little circles within circles that accounted for planets' strange movements across the sky. The planets moved in circles around the Earth and in circles in their little epicycles because circles were thought to be divine. "So they had a model that worked and it was descriptive, but it wasn't explanatory," Walker told me. "And it couldn't allow you to really say anything more than what the model described."[11]

Then came Copernicus. And Galileo, and Kepler, and Newton. And then, centuries later, Einstein, who while studying the behavior of light, understood that gravity came from the curvature of space-time. "All those little circles people were drawing in ancient times actually arose in the night sky because of this phenomenon we now call space-time. Which," she added, "is completely impenetrable to our everyday senses."

Walker thinks that our understanding of life, as a phenomenon, is right now where we were with gravity before Newton. We can describe what we see, but we have no sense of the underlying principles — we just see an apple falling to the ground. She thinks that without a theory, a deeper understanding of what life is, the search for it beyond Earth is preemptively doomed. And if we don't understand the phenomenon of life, how could we begin to mark its origin?

11 In a similar vein, Cleland has said that we are still at an alchemy-level of understanding life, not yet advanced into the true science of chemistry.

Collaborating with chemist Lee Cronin, Walker is working on a theory that could be the first step toward what she calls a physics of life. Cronin and Walker had each been troubled by the same gap in research, the assumptions made about what life is, and had been independently working toward a new framework that they each thought was needed. When they met, Cronin told me, "I didn't understand what she was talking about, she didn't understand what I was talking about" but, when expressed in math, their ideas turned out to be identical. In collaboration, they've come to call it *assembly theory,* a way of differentiating life from nonlife, not by its chemistry but by its complexity.

One way to assess a molecule's complexity is to smash it apart — science journalist Carl Zimmer likens it to breaking a Lego construction into its constituent bricks and counting them. Similarly, a big molecule can be smashed into its smallest components, simple molecules in this case. The more components a molecule has, the more steps were required in its assembly. And the process of those steps requires information, just as a massive Lego castle requires more information to be built than a simple tower.

Life's use of information allows it to do

things with matter that could not otherwise happen. Shake a bag of Lego bricks for long enough, and two might click together, but the Lego Death Star isn't going to be built on its own. "What that's trying to codify," Walker said, "is the fact that some things have so many steps to produce them that it requires a memory." That memory can be genetic or neurological or something we can't imagine, but the result takes too many steps to arise otherwise.

When Cronin tested the complexity of a range of molecules, some the result of biology and some not, he found a peculiar threshold. All the nonliving materials could be assembled in fewer than fifteen steps. Not every molecule in an organism takes more than fifteen steps to assemble, but nothing nonliving was made of molecules so complex.[12]

12 Fifteen didn't emerge from the data as a precise cutoff. Early in the research, Cronin was sharing his initial findings with an editor who, thinking of the most complex molecule not produced by life he knew, asked, "What about C60?" a soccer-ball style sphere of carbon. Cronin told me, "My algorithm massively overestimated how complex C60 was. And at that time the algorithm said C60 had

You can apply assembly theory to objects as well as molecules. A cell phone isn't alive, but in Walker's view it's still proof of life because it's so complex that life had to make it. Similarly, when Adamala says her synthetic cells aren't yet alive, she does say they're *biomarkers*. It's a word usually used in discussions of planets' atmospheres, the wisps and chemical whispers that could tell us life is there. But to Walker and Adamala, a biomarker is anything that could only be made by life's corralling of information. Walker picked up a glass mug on her desk. "Something like a cup doesn't appear in the universe without a living process."

We recognize a cup as a biomarker (even if you wouldn't use the word), but molecules aren't as obvious. And so Cronin avoids any approach that requires saying *Life is made of such and such molecules* and trying to recapitulate that origin in a lab — or even *Life is made of cells, so let me build them* — instead investigating the origin of life in an entirely

---

an assembly number of 14. So I just kind of said to that editor, 'Fuck you, 15.' Just make it one higher." Subsequent research showed that matter not created by biology peters out around 12, and living matter stretches all the way up to 30.

chemistry-agnostic way.[13] And he's more adamantly against historical recreation than any researcher I talked to. He asked me to imagine that, on Earth, we couldn't see the stars. (His hypothetical was we'd evolved a billion years later into the universe's lifetime, when expansion would have stretched the stars farther away, but you can also get there with an opaque atmosphere.) We would see our sun and wonder how it formed. "So everyone's obsessing about this single event that happened a few billion years ago, rather than saying, *How do suns form?*"

So rather than asking how life arose on Earth, Cronin is asking *How does life arise, period?* He has three of what he calls *chemical selection engines,* soups in the vein of more traditional origin-of-life research, but with no fealty to Earth's ancient chemistry. Each combines a permutation of simple chemicals — it doesn't matter that they match anything historical, but it matters

13 Some researchers think this approach is shifting the goalposts of the question of life's origin on Earth, but it can also be seen as asking a different question altogether. Not *How might life have arisen on Earth?* but *How does life, as a phenomenon, arise?* And then, *What can we learn about life, as a phenomenon from that?*

that they start simple, since complexity is what would signal an experiment's success. And each engine runs its seed chemicals through a series of environments, heating and cooling, inciting reactions — and tracking the evolution of complexity. Because that, Cronin believes, is the entirety of life.

Eventually, he's looking for the evolution of chemical systems that are self-perpetuating and create conditions that lead to more and more complexity. "You generate a universe in your flask," he said, "like a Big Bang. But it's not a big bang, it is literally that you've got chemicals reacting to each other." Atoms bouncing off one another and reacting, or not, seems random, but it all comes down to cause and effect. "If I flick, say, an atom to the left, it will go that way. And then if I flick an atom to the right, it will go that way. And as the atoms interact, although it's too complex for us to know why they're doing it right now, that is a genuine memory of those two things coming together." If those atoms bump together and form a molecule, that is memory. Some of those encounters turn out to be self-perpetuating, creating by their existence the conditions for their survival and for more of their kind to be made. If those molecules bump together and form something more complex, on and on,

Cronin said, "that is the selection equivalent of gravity. And that process of complexity generates everything we have in the universe that's associated with life."

Walker and Cronin's approach doesn't require any specific chemistry. And indeed, the search for life beyond Earth can seem painfully narrow, sometimes — searching for water, searching for carbon, searching for temperatures and conditions so like those we know here. There are two ways to justify this. First, that this chemistry may truly be the best chemistry for life. Second, that this is the only kind of life we'd know how to look for. You have to start somewhere.

Carbon is so intrinsic to life on Earth that it's synonymous — organic chemistry is the chemistry of carbon-based molecules. Not every molecule in your body, of course, is built on carbon, but all the big ones are. The complex molecules, the tricky molecules, the ones that allow life to do all the strange things that it does. And that's because carbon itself does strange and wonderful things. A carbon atom is relatively small, and thus lightweight, and can form up to four chemical bonds. Carbon also bonds remarkably well with itself, allowing it to be the backbone of long and complex chains, as well as other useful structures that branch or loop into rings. Carbon isn't just what life on Earth happens to be made of; it's exceptionally good at what it does, at what life uses it for.

Speculative alternatives replace carbon with silicon. In periodic-table terms, silicon is directly below carbon, meaning they have the same number of outer-shell electrons — four electrons available to form chemical bonds — but silicon is bigger, with one more full shell of electrons beneath those. And both elements are abundant — carbon more so in the universe, but silicon wins out in Earth's crust, and thus on similarly composed exoplanets.

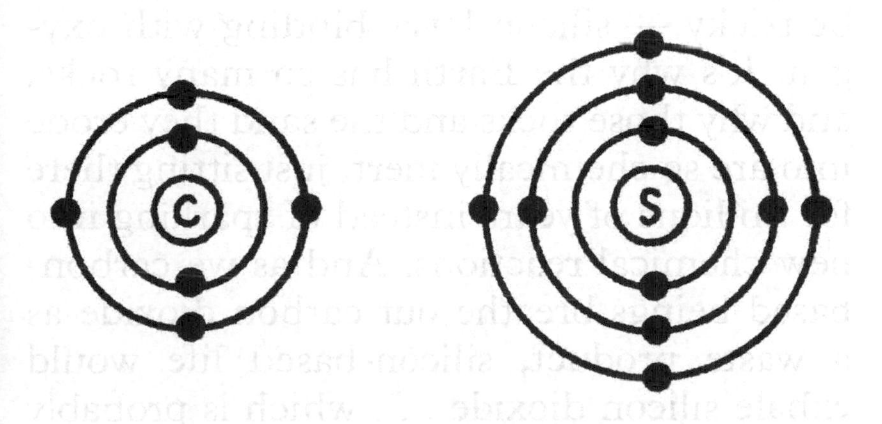

In one original-series episode of *Star Trek,* the *Enterprise* crew comes to a colony where miners are being killed by an unknown monster. The mine had recently been extended to new depth; then suddenly, the deaths, and the killer seems to be coming closer to the surface.

Mysteriously, no life-forms other than people have been detected in the mines, regardless of how thoroughly they scan. But those scanners rely on known chemistry. Spock muses, "Life as we know it is universally based on some combination of carbon compounds, but what if life exists based on another element? For instance, silicon." Kirk chimes in, "I've heard of the theoretical possibility of life based on silicon," but Bones, the doctor, scoffs: "Silicon-based life is physiologically impossible, especially in an oxygen atmosphere." And it would definitely be tricky — silicon loves binding with oxygen. It's why the Earth has so many rocks, and why those rocks and the sand they erode into are so chemically inert, just sitting there for millions of years instead of sparking into new chemical reactions. And as we carbon-based beings breathe out carbon dioxide as a waste product, silicon-based life would exhale silicon dioxide . . . which is probably most familiar to you as sand.

But the Horta — the silicon-based alien, which of course exists — primarily lives within rock, which, we're told, it moves through as easily as a human being moves through air.[14] When it appears on-screen with a blobby noise, it looks like a mound of green and brown rock laced with veins of some pizza-colored substance. It's about the size of — well, the size of a green and brown and pizza-colored rug thrown over a man scurrying on the ground on all fours. Before it is discovered to be without malice, the Horta will be injured by phaser fire. Bones goes off to save it and returns with his hands covered in wet concrete. The dying alien's chemistry being something like stone, Bones has spackled it to save its life.

But it turns out the doctor's initial skepticism was warranted, at least out here in the real world. Silicon, with its four bonding sites, can indeed form long molecules. But, as evolutionary biologist Mohamed Noor points out in his book, *Live Long and Evolve: What Star Trek Can Teach Us about Evolution, Genetics, and Life on Other Worlds,* "silicon binds more tightly to other elements like hydrogen or oxygen than to itself." Carbon atoms love to form long, complex chains,

14 Though, if that were how physics worked, you could walk through the solid body of an elephant for your shared chemistry. But it makes for a good monster!

but silicon-silicon molecules take little prodding, like just being dissolved in water, to disband and run off to form their preferential bonds.[15]

But for all the trouble of noncarbon life in water — could chemistry for life work without water at all?

We haven't found life yet on Mars (unless some really wild things have happened since I sent this book to the printer), but every conversation about that possibility hinges on the presence of water, either currently or in the Martian past. The very idea of a habitable zone, the ring of orbital space around a given star within which a planet could host life (as we understand it), is based solely on the ability for there to be liquid water on the surface. *How closed-minded,* I used to scoff, as a scoffy teen (and one who wanted

15 Noor does offer that silicon does have an option in an oxygenated atmosphere: "Life involving repeating units of silicon and oxygen ('silicones'), perhaps with carbon too, may work well in worlds with much higher temperatures than are typical on Earth — perhaps, in some respects, even better than our carbon-only-based forms." Organic material, after all, burns at far lower temperatures than do rocks.

life in the universe to be abundant). But water is a strange and unique molecule. It's polar, meaning its three-atom structure has a charge on each end; this makes it easy for things to dissolve in it — not just salt and sugar and other things you may stir into your pasta water or iced tea but all polar molecules.[16] In solid form, water has a crystal structure, which is (unusually, for a solid) less dense than its liquid counterpart. This allows for fun things like ice floating on the surface of a lake or ocean, which maintains winter environments for fish on Earth. Ice's buoyancy also means more surfaces, which are a thing life loves. Surfaces are the transition between two phases of matter — the ground is a meeting place of solid and gas; the sea is liquid and gas on its surface, and liquid and solid at its bed — and surfaces give life a place to be. An ice-covered body of water is a feast of surfaces, between the air and ice, ice and water, and water and solid ground.

One of the most talked-about possible

16 Substances we think of as *hydrophobic,* like oil, are generally so because their molecules aren't polar and so repel water instead of mixing in. As your high-school chemistry teacher may have said, *Like dissolves like.*

homes for life, right within our solar system, is Titan, Saturn's large and largest moon, where you'll find the only other open oceans around the sun. But they're not filled with water. They're oceans of methane and ethane, substances that are gases on Earth but are liquid at Titan's frigid surface temperatures (−290° Fahrenheit, or −180° Celsius, or so). And some scientists think they could be home to life.

Life on Titan would be as un-Earthlike as its environment. Planetary scientist Morgan Cable, who oversees NASA's Astrobiology and Ocean Worlds research group, told me that Titan's lakes are exciting largely because surface liquid is so rare in the solar system. Titan also has an underground ocean of liquid water, probably salty or shot through with ammonialike antifreeze, which is another possible home for life. But that underground, underwater life would be, chemically, Earthlike. Anything swimming in Titan's surface seas would almost surely be stranger, because it wouldn't live in water. Part of Cable's research entails investigating just what its chemistry might be. "It's challenging for anything to dissolve in methane and ethane," she said, "just for the polarity's sake — the fact that it has to be non-polar — but also because it's so damn cold." Her work to enumerate what might be dissolved in these lakes has so far resulted in a pretty short list: acetylene, butane, propane. Small, nonpolar molecules.

Could that be enough for life? Cable says she doesn't know. "We may be entirely surprised, right? We still think of life as entities like us." But, she reminded me, "There could be really weird things that, just because it's outside of our realm of experience, we haven't quite figured out how to frame the questions of what to look for."

When I asked Kate Adamala about life on Titan, she echoed Cable but in more skeptical terms. "Even if life in methane is possible, how would we even know it is life? We have absolutely no tools to investigate it, unless it's life enough to wave to us with a little hand." Barring hand-waving or tentacles, the only tools we have to find life use chemical signatures. We get hopeful if we see oxygen or methane, molecules that life makes and that don't last for long on their own, especially on a planetary scale. Microbes can leave fossils, too. But definitive life detection rests on chemistry we know from life on Earth. So life with different cellular machinery would slip through our net. Even the citizens of the proposed ancient RNA World would be undetectable by current searches.

This is where assembly theory could be key — Walker and Cronin propose disregarding the particulars, offering a way to recognize life not by its chemistry but by its complexity. Smash a bit of what you've found and see how many steps its creation took, if it was too much to exist without memory.

Life, to Cronin, isn't a sum of organisms, it's the inevitable result of the laws of the universe. "I think the process that drives the formation of life," he told me, "is as abundant as gravity is in the universe." Not every object exerts the same gravitational force — the sun is more massive than a pencil, so the sun seems to have abundant gravity while, to our experience, the pencil has none — but gravity is still inherent in all matter. It could be the same, Cronin says, for life. "I think evolution is an intrinsic feature of the universe, like gravity. It's just biology speeds it up."

Mike Russell, who first imagined life emerging in the microscopic pores of a system of undersea vents, also sees life's origin as just part of the universe's unfolding. In an article about Russell in *Aeon,* Tim Requarth wrote, "If you think of life in terms of energy, then life's emergence connects back to the very source of energy flow, the Big Bang itself." Energy and matter didn't dissipate after the Big Bang, didn't flatten out like a puddle into homogenous atomic soup, but clumped and hiccupped into structure. Stars formed, and planets, and surfaces, and seas. That disequilibrium eventually led to us as well.

From this vantage, the question of whether we're alone could almost become moot. We're not alone because we're not separate from the swirl of a galaxy's arms or the way wind catches dust in a gyre. We're no more an anomaly than an atom is. How could we ever consider ourselves alone?

But at the same time, life is also something apart from the rest. A protein is more than an atom, a cell is more than a protein — some thresholds are clearly being crossed. Even if the lines are arbitrary, the differences are not.

When we pursue knowledge about the origin of life, we're thinking about *what life is*. Is life self-replicating information? Is life a new way for the universe to organize energy? Is it, as Carl Sagan and others have put it, a way for the universe to experience — and hope to understand — itself?

It's all of those, of course. Life is information and energy and awareness. It's a squirreling away of entropy, so that one bit of ordered matter can look at another and try to know it. It's momentum rolling, for a moment, uphill.

Life seems sometimes like it should be impossible, and yet here we are, and yet here our whole planet is.

Understanding life's origin is a way to understand what it means to be alive, matter with memory, matter that brings new complexity into the world. And understanding how those origins — plural! — can happen helps us think about where else to look.

CHAPTER 2

PLANETS

About thirty years ago, Carl Sagan pointed two different cameras toward the Earth to see what we could see.

The first camera was strapped to *Voyager 1,* the probe that left Earth in 1977 for a tour of the outer solar system (also carrying with it the Golden Record, which bears messages for potential extraterrestrial listeners, and which we'll come back to in Chapter 6). In 1990, thirteen years after *Voyager*'s launch, the probe was out past Neptune, just under four billion miles away. Sagan directed the camera to turn back toward Earth and take one last picture. The resulting image, of a small, watery point against a vast black sky, became known as the "Pale Blue Dot."

Sagan wrote of it, in the book inspired by the photograph, "Look again at that dot. That's here. That's home. That's us. . . . Our posturings, our imagined self-importance, the delusion that we have some privileged position in the Universe, are challenged by this point of pale light. Our planet is a lonely speck in the great enveloping cosmic dark."

Sagan, whose work was so often aimed at a sense of cosmic oneness, hoped the image would actually make humans feel *alone:* "In our obscurity, in all this vastness, there is no hint that help will come from elsewhere to save us from ourselves." For all that he imagined extraterrestrial contact as a future turning point for humanity, he also hoped that awareness of our smallness in space could inspire humanity, beset by the threat of nuclear war, to save ourselves. The "Pale Blue Dot" became an icon of the environmental movement, alongside *Apollo 8*'s "Earthrise" image. The hope was that if we could see our world as a *planet,* we might understand its finitude and fragility and be moved to protect it.

But Sagan took a second, far less famous picture of Earth in 1990 from a probe named *Galileo.* Sagan confirmed the presence of water as gas, liquid, and ice; enough oxygen in the atmosphere that life was quite plausible; and, the loudest signal of all, modulated radio signals that could only conceivably be "generated by an intelligent form of life."

That life, of course, was us, observed during a flyby of the probe on its way to Jupiter, where it would go on to study the planet and its moons. Together, the "Pale Blue Dot" and the *Galileo* flyby inaugurated a research project that continues to this day: trying to look at Earth as a planet.

Thirty years ago, no one knew if there were planets beyond our solar system, but theories had abounded for centuries. In the 1800s, Kant's and Laplace's nebular hypothesis was popular, so planets were assumed to be common; in the early twentieth century, astronomer James Jeans proposed that planets were formed when two stars passed near each other and tidal forces stripped off some material that formed into planets; these flyby events would be random and rare, and thus planets would be, too. Around the 1950s, the idea of abundant worlds was once again popular, but while scientists could hope or imagine, they really had no idea.

And then, they knew. In 1992, the first planets were discovered outside our solar system, orbiting a pulsar, the rapidly spinning corpse of an exploded star — a seemingly inhospitable place for a planet to be, awash in X-rays and ultraviolet radiation.[17]

17 Though, some research has speculated that under the right conditions, those X-rays could make for a cozy planetary environment.

And then in 1995, the first exoplanet around a sunlike star was confirmed, 51 Pegasi b.[18] And then another, and another, first a trickle, then a flood. As of this writing, we've found 5,044.

Before the discovery of exoplanets, we looked to the planets in our own solar system to inform a model of planet formation, which nicely accounted for the worlds we knew, from tiny, hot Mercury to massive Jupiter and icy Uranus and Neptune farther out. It begins with the birth of a star, when a huge cloud of gas and dust collapses in on itself under the weight of its own gravity. As the star begins to burn, it sits within a spherical cloud of matter, which collapses into a spinning disk. Clumps and eddies in the disk coalesce as particles bump up against each other. The clumps get bigger and bigger until, long story short, they're planets. The planet's size and composition is determined by where in the debris disk it forms — closer to the star is warmer, so the available solid materials are bits of dust and rock, giving you small planets like the Earth and Mars. Farther out, in colder reaches, substances like water, methane, and carbon

18 Michel Mayor and Didier Queloz received the 2019 Nobel Prize in Physics for the discovery.

dioxide are frozen solid, making much more massive planets.

It's neat, it's elegant, and it got sort of shot to shreds once we started finding other planets. The basics were right, but the whole picture wasn't nearly so simple. The more we know of these worlds, the less we can take for granted.

Planets are another site to challenge our sense of specialness. Are planets like Earth common or rare? Rocky and wet, with enough gravity to hold an atmosphere but not so much as to crush a body, with seasons, with a moon . . . How long is the list of requirements for what we consider Earthlike, and are those the same as the requirements for life? Understanding the different kinds of planets that may bc out there leads us to ask: On how weird of a world could life flourish?

Holy Shit, That's a Planet

When I look at the night sky, I can point to the bright, steady lights and say, *There's a planet*. I say this often to my toddler, though he still responds, pre-Galilean, "Yes, that a star." I point to Venus, low in the sky after sunset, to Jupiter and Saturn, lately clinging to the sides of the moon. With a telescope, just like Galileo, I can even see Saturn's rings. And the first time I did, on a college rooftop observatory, I felt that sudden shock: Holy shit, that's a *planet*. Something I'd always known, now known in an entirely new way.

Exoplanets, on the other hand, orbiting other stars, are so distant and so small and dim that they can almost never be directly photographed. Scientists aim their telescopes and then infer and deduce. I've heard it described as trying to photograph a gnat flying around a floodlight — and both the gnat and the floodlight are at the opposite end of a football field from you. Only rarely and with great effort have exoplanets been directly imaged, little dots on photographs we can point to and say *Aha*. The planet orbiting 51 Pegasi b, the first found around a sunlike star, was discovered by measuring the star's *radial velocity*. Basically, a planet's gravity as it orbits the star tugs the star a bit into a wobbling, circular dance. The wobble is tiny, but it can be seen in variations of the star's color because of the Doppler effect — shading blue as it wobbles toward us and red as it leans away, its light waves either compressed or stretched.[19] (If the star's movement is lined

19 Just as sound waves are subject to the Doppler effect — a siren going up in pitch as an ambulance approaches you, then down as it drives

up right to our view, the wobble can also be seen in the star's position in the sky, a detection method called *astrometry*.)

Scientists also find planets by the tiny eclipses they make when they pass between us and their stars. The planets are too small, compared to their stars, to come close to fully blocking out their light, but they do slightly dim our view of the star when they pass in front of it. These dips in a star's brightness, the line called its light curve, create a signature called a transit.

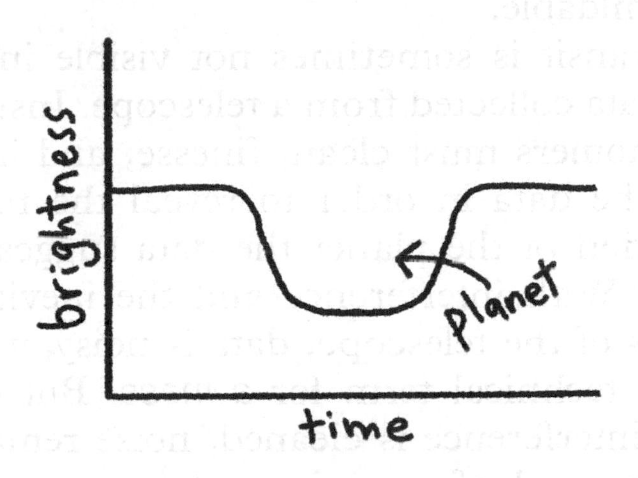

away — so too are light waves compressed or stretched. But instead of pitch, this change is experienced in color, objects looking redder as they move away from you, bluer as they move toward you. You can't notice this in the color of a speeding ambulance, but for a star or galaxy, with a sensitive telescope you can.

A transit is as close as we can come to seeing these planets, but the curve's decisive shape is, if you know to understand it, as clear as a picture of the sky. Anthropologist Lisa Messeri embedded with teams of astronomers to write *Placing Outer Space: An Earthly Ethnography of Other Worlds*. Observing (and sometimes helping carry out) their research, she saw how astronomers learn to see a planet in a light curve, and learn to see a transit in far less decisive data. The interpretation — and imagination — in that work is formidable.

A transit is sometimes not visible in the raw data collected from a telescope. Instead, astronomers must clean, finesse, and interpret the data in order to reveal the representation of the planet the data suggests is there. With interference and the inevitable quirks of the telescope, data is noisy, which is the technical term for *a mess*. But even after interference is cleaned, noise remains, and the work of astronomers becomes markedly subjective. In one scene in her book, Messeri describes an astronomer new to this sort of research — a theorist inexperienced in working with raw data — learning from two astronomers who do the work regularly. The two experienced researchers use different filters for processing the data. When the theorist asks one astronomer why she uses the filter she does, it's simply because "a former boss introduced her to it." The choice wasn't haphazard, but it also wasn't the only one that could be made.

As Messeri sees it, astronomers make up for the lack of data-packed photographs of exoplanets by constructing a suite of "abstract representations," light curves and data plots that, within the field of study, are how astronomers turn data into a planet and a planet into a place. She cites other scholars who describe the work as an aesthetic project — "an aesthetic not of beauty but of realism." Worlds made real.

"Place," Messeri writes, "suggests an intimacy that can scale down the cosmos to the level of human experience." It is what distinguishes the study of exoplanets from that of other cosmic objects — galaxies and black holes and stars.[20] The cosmos, for all its

20 I think we might be close to making asteroids places, with the recent OSIRIS-REx mission that touched down on the asteroid Bennu to take a sample to bring back to Earth. Something about the way rovers and space robots are personified — once one has been to a heavenly body, that body becomes a place where we can imagine ourselves, too.

vastness, becomes speckled with islands of familiar scale, rest stops where we can catch our imaginative breath. And finding that familiarity out in space is a thrill. "Places are exciting," Messeri writes, "because we know how to know them; we all have experience learning what it means to be somewhere." And we know what sorts of questions to ask about them.

When I asked astronomer Jessie Christiansen, science lead at the NASA Exoplanet Archive, what's happening inside her imagination when she's thinking about an exoplanet, she told me she basically outsources that work to collaborators, artists who create the images that go along with press releases announcing the discovery of new exoplanets. Christiansen said, "I bring them the numbers — it's this big, it weighs this much, and the star is this temperature. And then . . . they make something." These artists are the ones who turn the place an astronomer can understand into one the public can imagine their way onto.

So I asked the artists how they do that.

Collaborators Tim Pyle and Robert Hurt came to the work from opposite poles. Hurt is an astronomer who had always dabbled in graphic design and art, and Pyle worked for years in visual effects in Hollywood and applied for his current job at Caltech with an animated short about bees taking over some NASA equipment and exploring space. Their work is far broader than creating art to accompany press releases — they craft visuals and videos and multimedia experiences, any way of "telling the science story" — but their illustrations of exoplanets are no small part of their project. These images, which often accompany news stories about the newly discovered exoplanets, have to draw readers in, to set up their framework for the story they're about to read. Pyle says, "It has to be attractive on some level — in the sense of being eye-catching . . . It can be a very ugly planet that's attractive," a cosmic *jolie laide*. Hurt says, "We need to show them a picture that's at least starting to be as cool as what you saw on the *Guardians of the Galaxy* trailer, if we want people to stop and linger and wonder what it was."

Pyle and Hurt — and other artists like them — create vivid, evocative images with a very few pieces of data. As Christiansen said, all that's known is "it's this big, it weighs this much, and the star is this temperature." Pyle said that everything more than that is "educated likelihoods." Hurt framed the artists' decisions as part of the scientific process. "In the best case scenario, [it] should be viewed as a hypothesis," he said, a foray or a positing, a few steps out on a limb. "We have to overcommit what we know."

But fidelity is not the artists' main goal. Instead, the artists and scientists have to overlay the decisive data with reflections of human ideas. For example, we have no way to definitively know if a planet in a certain range of sizes is a super-Earth, rocky and wet and big; or a mini-Neptune, with a thick, swirling atmosphere of hydrogen, helium, and methane gas enveloping a slushy mantle of frozen water and ammonia. That's a particularly tricky kind of planet, at the transition between large rocky planets and small ice giants, but every example has its unknowns. To turn the data into an illustration, an artist has to choose — not just what the data suggests but what sort of illustration serves the situation.

And, indeed, Pyle told me, "I know of at least one person who is actually of the mindset that because there could be these multiple routes you could go, we should not even be attempting to illustrate exoplanets," because in doing so, the artists and scientists make choices based on factors other than data — on aesthetic preference or narrative power or something even less tangible. "We know that whenever we do an artist's concept, it will be wrong."

One of Pyle's illustrations is of a planet called Kepler-452b. At the time of its discovery in 2015, it was the smallest exoplanet that had been found in its star's habitable zone, about sixty percent larger in diameter than Earth — and, warranting excitement and press releases, it was in its star's habitable zone. Roughly.

Based on the planet's size and location, it was possible that if this planet had an atmosphere, it could be like Venus, with a thick haze of clouds — scientifically meaningful, but rendering it in art would obscure the planet's presumed rocky surface. But it was also possible this planet wasn't a hothouse at all, just a slightly warmer, slightly bigger Earth. So Pyle split the difference, illustrating a planet with gray-brown continents and greenish-blue seas under hazy clouds. But the coastlines are shaded lighter, and riverbeds are crusted with salty white. There used to be more water here, the art suggests, but the planet is heating up, the rivers and oceans evaporating as the atmosphere fattens itself with water vapor, a thickening and thickening blanket.

I wouldn't have noticed those salty shores, let alone thought through their implications, without Pyle's guidance. I'd have seen a rocky planet with a little bit of water, less inviting than Earth. And Pyle's version gives us a far less specific scene than one from the SETI Institute. Their image, drawn not from the point of view of space but from right over the surface as if the viewer is in flight, is filled with golden fjords rising out of deep blue rivers, and volcanoes clouding the skies. In the foreground, thin fingers of rock jut out of the water. The caption here says, "This illustration imagines that a runaway greenhouse effect has begun to take hold on Kepler-452b, driving off much of the planet's surface water." The same story in a vastly different scene.

Most of the star systems we've found look nothing like ours. Partly, that's due to the methods we've had for detection. The Kepler space telescope, which searched for exoplanets (via the light curve/transit method) from 2009 to 2018 and is responsible for roughly half of the planets currently known, was most sensitive to large planets with tighter orbits around smaller stars. But it does seem like our solar system may be strange.

In line with the Copernican principle, the base assumption before we'd found any exoplanets was that our solar system would turn out to be average. But when scientists started finding exoplanets, their elegant model was shown to be far too clean. That first planet found around a sunlike star, 51 Pegasi b, is 150 times more massive than the Earth — about half the mass of Jupiter — but it orbits closer to its star than Mercury orbits the sun. More of these hot Jupiters were found soon after. They're not particularly common, it turns out — they account for maybe 1 percent of planets — but being huge and having very short orbits, they were the easiest and fastest to find.

Nowadays the best guess for planet formation is a more chaotic story (and much less subject to scientific agreement). Planets are still thought to form out of the debris disk around a new star, smaller rocky planets close in to the star, gas and ice giants farther out. But now scientists believe there is much more orbital tumult before planets settle into their clockwork dance. Gas giants may spiral in toward their stars, destroying or ejecting some or all of the smaller planets. The first batch of rocky planets may fall into the star, with longevity left to a second generation. Even in our seemingly orderly solar system, some scientists think Jupiter may have migrated inward only to be pulled back out by Saturn or other forces unknown. Other discoveries have added further complications. Very large planets orbiting very far from their stars suggest an entirely different birth story, forming not by accretion — dust to pebble to rock to world — but from gravitational instabilities in the disk. As astronomer Scott Gaudi put it to me, joking but not really, we know less about exoplanets now than we did before we started discovering them.

Our solar system is well populated and diverse, but exoplanets turned out to be more exotic. There are hot Jupiters and super-Earths and mini-Neptunes. (The super-Earths and mini-Neptunes create their own host of problems for our theories of formation, as they're in precisely the range of masses the original formation theory predicts very few planets should be, yet they seem to be the most common type out there.) There are some planets wandering lonesome through interstellar space, dark and cold and starless, probably ejected from their home system by the gravitational shenanigans of a migrating gas giant. Jessie Christiansen told me that her favorite system is K2-138, a suite of six planets orbiting a star a bit smaller and dimmer than our sun, about 800 light-years away. The planets are all in the super-Earth/mini-Neptune range, but they're packed in just about as close to the star as possible (any closer and their gravity would perturb each other out of their orbits). And their orbits are all in resonance: for every three orbits a planet makes, the next one out makes two. "It plays music," Christiansen told me.[21]

The planets of K2-138 were found by the transit method, the shadows of these worlds temporarily dimming our view of their star. That method is best at detecting planets that orbit close to their stars, because in order to confirm that you've found a planet, you ideally want to observe several transits. The closer a planet is to its star, the shorter

21 At the website System Sounds, a team of musicians and astrophysicists turns astronomical data into music — sonicalizations — where we're used to data being visualized. Really, any system can be turned into music. The method here is to assign each planet a note, translating their orbital frequency to a frequency in the range of human hearing. The 3:2 resonance of K2-138's planets means that each pair is, musically, a perfect fifth apart. So, according to System Sounds, "these five planets follow the tuning system designed by Pythagoras himself in which all notes are related by a perfect fifth." Each time a planet completes an orbit, its note plays. Any system could be converted into notes, but K2-138 is one of the rare cases where the sonicalization is musical.

its orbit, which means you can catch more transits in a given time frame. Mercury, the closest planet to our sun, completes an orbit every 88 days; Earth takes 365, as you know, and distant Neptune's orbit takes 165 Earth years. So finding Neptune by the transit method could take a millennium.

Thus planets found by the transit method tend to have smaller orbits. The radial velocity method, which relies on stars' Doppler wobble, is good at detecting big planets, which have (relatively) dramatic effects on their host stars. Together, these methods have given us a catalog with a big blind spot where Earth would be — we see the very big planets and the tightly orbiting planets, but beyond that it's mostly extrapolation. And that extrapolation, while conscientious, is contentious. Different scientists, using different but totally reasonable math, come up with a wild range of predictions for how common Earthlike planets are: some say 2 percent of sunlike stars have Earths, some say basically all do.

The desire for Earthlike planets hums along under all the planet-hunting. With each new discovery, our models of planet formation evolve, and our understanding of cosmic diversity expands. But strange planets also thwart a deep human desire.

The search for Earthlike planets is partly the search for homes for alien life — there's good enough reason to think life needs liquid water, as well as the sorts of surfaces a gas giant can't provide. We're also, on some level, seeking new homes for humanity, the search for our future among the stars. But even if our descendants never leave Earth, a cosmos full of Earthlike worlds is one that is, wholly, our home.

EARTH ET AL.

Fiction writers take Messeri's place-making even further, of course, making planets not just places but settings, their quirks and constraints driving the narratives of characters' lives. In Stephen Baxter's novels *Flood* and *Ark,* humans flee an Earth catastrophically flooding (from tectonic activity, not [just] climate change). The crew fractures en route, disembarking at (at least) two planets. They name them Earth II and Earth III; the novellas set on those worlds 400 years after landfall have the same names. And in each, Baxter extrapolates from what can be known about a strange, un-Earthlike world to imagine how humans would make a home on its surface.

Earth II is Earthlike but tipped on its side, rotating as Uranus does with its axis parallel to its orbit. (Planetary scientist Lindy

Elkins-Tanton explained it to me, when describing the similarly sideways asteroid Psyche, as a rotisserie chicken.) A planet thus rolling as it makes its way around its star, rather than spinning like a top, would have the most divergent seasons possible.

Earth's seasons come about because of the gentle tilt of the planet's axis, currently 23.5 degrees. In summer the northern hemisphere is tilted toward the sun, and in winter it's tilted away, and so warmth comes and goes accordingly. A perfectly vertical planet would have no seasons. A sideways roller like Uranus or Earth II would have the most.

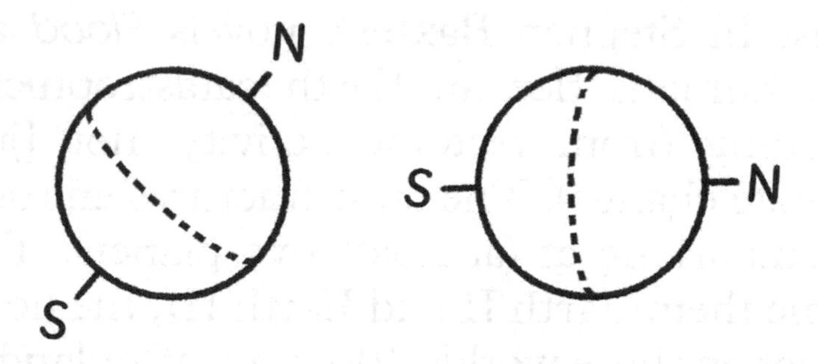

At the solstices, rather than being slightly tipped toward the sun, Uranus points a pole at it just about straight on. The sun is directly over the north pole in winter, over the south pole in summer. And so not only do the polar regions have periods of perpetual night and day but so does each entire hemisphere.

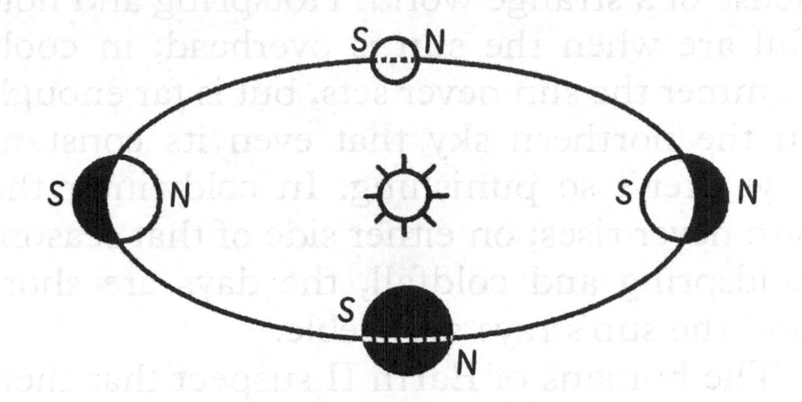

On Earth, the middle latitudes never see the sun straight overhead, but on an otherwise Earthlike rotisserie planet, you'd get that true high noon a few weeks after the spring solstice. And as the days moved toward summer, the sun would move farther and farther north in the sky — the days would continue to lengthen, more than twelve hours of daylight at a time — but the sun's rays would be more slanted, so the heat would actually abate, even through a few months of unsetting sun. And then in autumn there would be another spike of heat when the sun was directly overhead, and then for the whole hemisphere several weeks of endless night in winter.

Baxter imagines his humans in a world of six seasons: coldspring, hotspring, coolsummer, hotfall, coldfall, coldwinter, Earthly season names coopted to make sense of a strange world. Hotspring and hotfall are when the sun is overhead; in coolsummer the sun never sets, but is far enough in the northern sky that even its constant rays aren't so punishing. In coldwinter the sun never rises; on either side of that season, coldspring and coldfall, the days are short and the sun's rays are feeble.

The humans of Earth II suspect that their planet once rotated upright as most planets do, but, as is believed to have happened to our own Uranus, a massive impact at some point toppled it over. But Earth II's scholars discover the planet's tumultuous days are not behind it: their planet, they realize, periodically or occasionally, wobbles on its axis. Perturbed by two giant planets farther out in its system, Earth II's axis will tip up a bit out of its plane of orbit, disturbing the seasons and churning up volcanic activity. (It's a bit like the movie *2012,* but . . . vaguely plausible.) The wobble, the scholars fear, will change the planet's seasons. Midlatitudes will get milder seasons, but the poles will no longer get their head-on sun. If ice caps form at the poles, then sea levels will drop — perhaps not as cataclysmic as rising waters inundating coastal cities, but disruptive nonetheless. And as water redistributes itself around the planet, storms, droughts, and floods may come. The wobble may also, in a sense, shake the planet up literally, too. "And even worse than that, the tipping planet will shudder. There will be earthquakes and volcanoes." The scholars don't expect it to be the end of humanity, but perhaps of civilization. "Buildings would crumble, the seas would rise, volcanoes would fire." Life on such a capricious planet would be hard-pressed to attain any cultural longevity.

But even the looming planetary disruption aside, humans on Earth II feel that something's just not right. One Earth II–ling remarks, "The seasons change so quickly around the equinox, from coldspring to the torrid heat of hotspring, just a few weeks. It always seems to catch me by surprise." His tutor responds that it should indeed be surprising: "We humans evolved as tropical animals on Earth — a planet, I remind you, of moderate seasonality. It's said that even on Earth those who lived at high latitudes felt surprised every year at the abrupt changes of length of the day."

Elsewhere, the humans of Earth III are similarly disoriented by their adopted home's strangeness, too static in this case rather than too variable. Earth III orbits a smaller, dimmer star than the sun, which means that in order for it to be temperate, it must be much closer to its star. Planets in very close orbits tend to fall into spin–orbit resonance with their star — in the most extreme case, tidally locked, so the planet makes one spin on its axis for every orbit around the star. The moon is tidally locked with the Earth; this is why you only ever see its one face.

The moon still has phases because it doesn't always show the same hemisphere to the *sun*. For a body tidally locked to its star, though, one side gets blasted by starlight while the other gets none. If that planet had no atmosphere, one side would bake and the other freeze. Even Mercury, not in a one-to-one resonance but with such a slow spin that a single day there lasts as long as 58 days on Earth, has daytime surface temperatures peaking around 800° Fahrenheit (400°C). At night, its temperatures can drop to −290°F (–180°C). But those swings aren't due just to the length of day and night, but to the lack of an atmosphere that would moderate temperature, holding heat at night like a blanket and potentially reflecting back some of the sun's energy during the day. An atmosphere also distributes heat through wind and weather — the stark differences between a permanent day and night side might drive powerful winds, but those winds would use up and transfer some of the hot side's excess energy.

On Earth III, Baxter imagines a gentler possibility, a temperate climate on the sunward side, where climate zones, instead of striated north to south as on Earth, ring the substellar point like a bull's-eye. And there, too, the human settlers are unsettled, in this case by their sun that hangs motionless in the sky. A soldier laments, "It doesn't change. . . . Watch after watch. The Star just hangs there in the sky, and we all sit in the mud, waiting." The idea that humanity came to this planet from "a turning world" is contentious, but, his companion muses, "I wonder if we miss that, on some deep level."

We see these planets in opposition to Earth — harsher, disconcerting, strange because they're different. Every un-Earthlike world we can imagine inhabiting becomes a more challenging place to live — the tumultuous seasons and terrible wobble of Earth II, the cold darkness of Earth III's back side. Earth sometimes seems like the most habitable planet possible — not just for us but possibly at all.

The ideas of Earthlikeness and habitability blur. It's hard to think of the important things that make Earth as separate from what makes it able to support life. Lisa Messeri writes, "A *habitable planet* is not a natural thing simply existing out there in the universe, waiting to be discovered; *habitability* must be imagined, defined, and made important. For the exoplanet community, *habitability,* a rather unglamorous word, has become shorthand for what astronomers consider the greatest discovery their field, and possibly humanity, can make." Habitability is a measure of a few simple factors; it is also the promise of life, the promise of planets that are not just places or worlds but possibly homes.

The most basic criteria is composition: an Earthlike planet should be rocky, with a surface that gives life somewhere to be. Rocky worlds are necessarily of a certain size — if a planet forms much more massively than Earth, its gravity holds on to a thick gaseous envelope, resulting in something more like Neptune or Saturn. A planet in the super-Earth/mini-Neptune size-and-mass range, even if it has a rocky surface and an atmosphere, may have so much gravity that it compresses its atmosphere too densely, at too much pressure, for liquid water to exist on the surface at all.

A habitable planet should also be in the

"habitable" zone around its star. A bit confusingly, that location doesn't guarantee habitability. The habitable zone is the distance a planet's orbit could be from its star to have the right surface temperatures for liquid water. Not too hot, not too cold, it's also called the Goldilocks Zone. Abel Méndez, professor of physics and astrobiology at the University of Puerto Rico at Arecibo, maintains a catalog of habitable exoplanets. "The criteria is basically the size and the orbit," he told me. "That's it." The orbit accounts for the temperature, and the size of the planet dictates if its surface can hold liquid water. Méndez said that a planet with 1.5 times Earth's radius would have about 1 percent of the planet's mass as atmosphere. Earth's atmosphere, in comparison, is one-billionth of its mass. "At the surface level the pressure will be so high," he said, "that it will make any ocean, any liquid water, solid." On the other hand, if a planet is too small, its weak gravity means it will have no atmosphere or one too thin, and all the water will sublimate and escape into space. This is why Mars, half the size of Earth, may have once had water but lost it.

When Méndez started keeping a catalog of potentially habitable worlds in 2011, he said, there were only two or three candidates. "I thought, maybe this is a ridiculous idea." He laughed delightedly, "I was wrong!" First there were two or three, then six, then twelve. "I knew all the names," he said. "And then fifty? Forget it. Okay, we need a catalog."

The Constant Moon

But what else makes the Earth *Earth,* beyond the planet's size and distance from its star? Is it our moon, stabilizing our orbit and thus seasons, and whose formation may have thinned the Earth's crust enough to facilitate the plate tectonics that may drive evolution and stabilize the climate by recycling carbon from the atmosphere into the planet's rocks? Well, that sounds pretty convincing.

Lots of planets have moons, but ours is decisively larger than most, in proportion to its host planet.[22] Most moons form the way planets form, in microcosm. They coalesce from the debris left orbiting a planet after its formation. Our moon had a much more violent birth. The going theory for its origin is that late in the solar system's formation, about 4.4 billion years ago, a Mars-sized proto-planet that scientists call Theia slammed into the Earth. (I encourage you to

22 From what we can see in the solar system.

go online and find a video simulation of this event. It's very cool.) Theia slammed into the Earth, destroying itself and liquefying most of the Earth's surface. Much of the material ejected in the collision spiraled back to the Earth's ravaged surface, such as it was, eventually to resolidify. But over the course of — fittingly — about one month, some of it coalesced into what we now know as our familiar and surprisingly large moon.

The moon, thus, is made of Earth, a powerful discovery only realized when *Apollo* astronauts brought moon rocks back with them to be inspected. Our satellite's heft holds the tilt of Earth's axis stable, at the nice 23.5 degrees that gives us the variety of seasons without catastrophic swings. And while the Earth's axis wobbles a little, ranging by a few degrees over the eons, planets without big moons have wobbles that verge on flops. Mars, for example, is believed to wobble 10 degrees in either direction from its current 25-degree tilt.

The moon also gives us our tides, stronger and more variable than they would be from the sun alone. And tides may have shaped evolution on Earth, Rebecca Boyle writes in *Scientific American,* "shepherding the first plants and tetrapods [four-legged animals] from the salty marshes of the coasts and onto land." The ability to survive in tidal zones, sometimes submerged and sometimes dry, encouraged sea creatures to evolve toward land-dwelling.

We may also have the moon to thank for plate tectonics. Theia's impact stripped away some of the Earth's crust and mantle, and as Theia's core sank to the center of the reforming planet, the blasted-off material coalesced into the moon, leaving the Earth with a thinner crust than it would have had otherwise. With Earth's original, thicker crust, the heat of the mantle wouldn't penetrate, and our planet would ossify and stagnate. Venus and Mars have no plate tectonics; our violently formed moon could be what makes the difference. (Another theory is that as the moon has gradually receded from the Earth, as indeed it is very slowly spiraling away, it exerted tidal forces on the planet's body that triggered tectonic activity.)

The moon's influence seems to keep the Earth in a sweet spot of stability — neither so static that life is never challenged to evolve nor so changeable that the environment becomes hostile. And in that balance, Earth seems not just habitable but practically perfect in its way, every environment a home to some sort of life, the whole world densely and diversely inhabited.

But the planet didn't come preset with a perfect environment. Life's presence shapes the planet, too. Early life pumped the atmosphere with oxygen, which enabled the rise of more complex, energy-hungry beings. It's a snowballing of habitability, life's processes remaking the planet into a home. Everything about Earth seems tailor-made to Earth life, but part of that is Earth life's doing.

Now humans are well on our way to making this most habitable world inhospitable. The planet itself will be fine, with its rock and water and thin atmospheric blanket. It's ourselves we're going to ruin things for. Not the end of the planet, but the end of the world.

"Let's start with the end of the world, why don't we?" begins the prologue of N. K. Jemisin's *The Fifth Season.* "Get it over with and move on to more interesting things." The *more interesting things* here are how the end of the world is survived, how people move from "the ending of one story" to "the beginning of another."

The titular planet of Jemisin's *Broken Earth* trilogy is an angry one, whose people have had to build their culture around surviving its whims. They call its outbursts *fifth seasons*. A fifth season is a time of geologic upheaval, when earthquakes and volcanoes bring a new kind of winter, one of ash and cold, acid rain and starvation. A volcano erupts and darkens the sky, or acidifies the jet stream, or vaporizes the contents of a great lake, filling the atmosphere with steam and dust. People, here, have learned to survive through principles passed down like folklore. When a season is declared, states and municipalities collapse down to walled settlements in a kind of survivalist martial law. Seasons only come every few hundred years, but the core of society is about surviving them.

"[N]o one speaks of celestial objects, though the skies are as crowded and busy here as anywhere else in the universe. This is largely because so much of the people's attention is directed toward the ground, not the sky. They notice what's there: stars and the sun and the occasional comet or falling star. They do not notice what's missing."

What's missing is a moon. This roiling world is one that once had one, but their moon has been lost.

We eventually learn that this moon was flung out of its orbit in an act of human hubris, an attempt to plumb the deep energy stores of the planet. Now, the moon wings around on a wildly elliptical orbit, perhaps triggering seasons with its cyclical perigee. But without the moon's stabilizing influence, the planet is a seething wreck. Life clings on, humanity subsisting by its wits, plants and animals by lucky adaptations. One placid animal turns vicious and carnivorous in a fifth season; plants hibernate or start producing poison. The planet becomes not a place or a home but an antagonist. Jemisin says the seed of the story came to her in a dream, an image of a woman walking with an entire mountain floating behind her. *What made this woman so angry that she could move a mountain?* the author wondered and set out to discover in writing the books. But the world turned out to be angry, too.

Tectonic activity can be catastrophic, but at less cataclysmic scales it makes the planet itself seem alive. A volcano can level a city, but it also births new islands and releases gases from the mantle into the atmosphere, carbon dioxide and water vapor along with the more noxious sulfur dioxide. Earth lost the hydrogen atmosphere it was born with relatively soon after it formed, hydrogen being the lightest element and Earth's gravity not enough to hold it; volcanoes give a planet of Earth's size a second chance at air.

Plate tectonics — volcanic eruptions and the recycling of the crust — are also a crucial part of the global carbon–silicate cycle, which regulates Earth's temperature by increasing carbon dioxide in the atmosphere when the planet cools, or removing carbon dioxide from the atmosphere when things get too hot. The more CO_2 in the atmosphere, the warmer the planet; the warmer the planet, the more acid rain; the more acid rain, the more weathering of rock; the more weathering of rock, the more carbon runs off into the oceans, available for little sea creatures to use to build their shells; the more little sea creatures build their shells, the more carbon is trapped in shells and seafloor sediment when those little sea creatures die and sink to the bottom, to eventually be churned back into the Earth's mantle where tectonic plates meet and one slides beneath the other.

Pulling that CO_2 out of the atmosphere and sticking it under the ocean lets the planet cool off. As the planet cools, there's less acid rain and the weathering slows, and less CO_2 is pulled out of the atmosphere, while the gas is replenished from volcanic emissions. The resulting greenhouse blanket warms the planet back from the icy brink. That cycle seems to hold the Earth in its habitability — not so strongly that humanity can't mess it up — and without plate tectonics, the subduction of crust under the oceans, and the expulsion of fresh matter from volcanoes, it probably wouldn't work.

When we're looking for Earthlike planets, it's tempting to carry a long list of criteria, all of Earth's particular qualities, like plate tectonics, that we see are important for life. But the universe could just as well have come up with other solutions.

Only about 7 percent of stars in the galaxy are like our sun. There are far more red dwarfs out there — small, cool stars that burn for trillions of years longer than their hotter cousins. And odds are, most of them have planets (Earth III would be one of them). Their habitable zones can be measured, too, but while you'd be warm enough for water there on a close orbit, these stars are more active than our sun. Do their intense ultraviolet rays keep any planets near them sterile, or does life, you know, find a way? There are also planets around white dwarfs, the simmering cores left when dying stars bloat into red giants, as the sun will do in five billion years, expanding to engulf even the Earth. Did some planets somehow survive that blast, or did they form later, out of the rubble? White dwarfs persist for tens of billions of years. Maybe that's enough time for life to figure something out.

Moons themselves can be habitable, too. Instead of depending on a star's energy, a moon could be heated by internal radioactivity or the tidal pull of its planet, so standard habitable-zone rules may not apply. And while our large moon seems unusual in size and circumstance of birth, it's only big in proportion to the Earth; a giant planet's proportional moons could easily be big enough to support life. Astronomer David Kipping told me, "Our moon obviously isn't habitable, but it's in the habitable zone. And if it were a bit bigger — if it were the size of Mars — it would probably be a habitable world in its own right." The moon has 1.2 percent the mass of Earth; get closer to 10 percent, he said, which is more like the proportional sizes of Pluto and its moon Charon, "then I think there's a very good chance that Neil Armstrong would have been stepping out in a T-shirt and shorts. It's wild to think how different human history would have been if we'd known there was a habitable world right above us."

Fiction takes advantage of this real estate — in the moon-world of Pandora in *Avatar*; or in Ursula K. Le Guin's twinned worlds of Urras and Anarres in *The Dispossessed*, a planet's very large satellite proving convenient when anarchist revolutionaries need a new home to form their splinter state. Kipping is searching for these moons' real-world counterparts, using some of the same techniques that planet-hunters employ. Exo-moons may be habitable worlds themselves, or they may make life on their planets possible, stabilizing their orbits and quickening their crusts. We hold out hope that the galaxy is inventive.

2,000 Imagined Observers

When I interviewed Abel Méndez, who oversees the catalog of habitable planets and whose research investigates the boundaries of habitability on other worlds, I asked him what the idea of finding life beyond Earth meant to him. He laughed and said, "The short answer is, I don't even care anymore!" But he anticipated my next question and posed it himself: *"Well, so why are you still doing this?"*

The more Méndez learned about life on Earth, the more he saw how special it was — and, potentially, how rare. "You realize how special Earth is in this process. And so my thinking was now, well, extraterrestrial life is not that interesting to me . . . Home is your most interesting thing." So why keep looking to the skies? "Because I learn more. From studying all this, I learned more about Earth. And that's what I love."

To see an alien world as habitable and a habitable world as alien is both destabilizing and a source of connection. Research that looks at Earth as an exoplanet is technically about testing methods by which scientists might study other worlds. Astronomer Enric Pallé told me, "What we're trying to understand from looking at the Earth as a planet is to know which are the observables that we can search for in exoplanets, [observables] we see on the Earth that can give us clues about what these planets are made of, and what are their atmospheres made of, and even try to infer if there is life on the surface."

Most studies that imagine Earth as an exoplanet do so by reducing close-up images from a satellite or spacecraft to an extreme zoom-out. Messeri writes of turning planets into worlds, but these studies are turning a world into a planet. You can do this by compressing a satellite image into a single pixel and tracking how those images change over time, or by studying sunlight that's passed though the Earth's atmosphere as it reflects off the moon, which makes a good stand-in for starlight filtering through an exoplanet's atmosphere. Research using these methods has shown that a distant observer could detect the abundance of photosynthesis, measure Earth's rotation, and even determine the general layout of our continents and oceans.

But the question *What could a distant observer see of Earth?* contains within it another, more exciting question: *Who are we imagining doing the observing?*

Inherently, it's someone who wonders if we are here.

All of these approaches imagine a distant observer at some unspecified place, far enough away from the Earth to see it as a single point. Lisa Messeri calls this imagined distant location an Archimedean point — that is, a point far enough away to offer a different, perhaps more objective or truer, perspective. (The philosophical term comes from Archimedes's statement, "Give me a lever long enough and a fulcrum on which to place it, and I shall move the world.") But sometimes, the distant observer is more explicitly imagined.

In research that was published in 2021, astronomer Lisa Kaltenegger (fittingly, director of the Carl Sagan Institute at Cornell) led a team using new data from the European Space Agency's Gaia observatory to figure out which stars currently have, or have had, or will in the next 5,000 years have the right view to detect Earth in orbit around the sun, using the same technology that we use today to detect planets around other stars.

Kaltenegger told me that part of her inspiration — aside from the new availability of this Gaia data — was to help guide future searches for extraterrestrial intelligence. But in a way, she's manifesting that extraterrestrial presence by the very nature of her study. Her work doesn't imagine Earth observed from an unspecified distant location. It proposes a real viewpoint — 2,000 of them, in fact. As Messeri put it when I asked her about this research, "The viewer is emplaced on a star." Or, to be plausible, on a planet orbiting a star. Messeri writes in *Placing Outer Space,* "Scientific practices of place-making turn the infinite geography of the cosmos into a theater dotted with potentially meaningful places that are stages for imaginations and aspirations." But research that makes Earth an exoplanet does more than let us imagine ourselves someday inhabiting a distant planet — it evokes a distant planet that's inhabited by someone else. And they're looking at us.

With the Gaia data, Kaltenegger was able to deduce not only which star systems can detect the Earth's transits right now but which have been able to see us for the past 5,000 years and who will for the next 5,000.

Ross 128, a red dwarf star about 11 light-years from us, known to have a planet just a bit more massive than Earth orbiting in its habitable zone, first became able to see Earth's transit 3,000 years ago but lost that view around the year 1100 CE. We're no longer in the perfect edge-on orientation for them to detect Earth via the transit method, but there are other ways to detect life, including radio waves, which don't require such precision. Kaltenegger told me, "I started to wonder, if they'd seen a planet [a couple thousand years ago] and now all of a sudden, if there's radio waves from the same part of the sky, would they put it together?" Would a Rossian astronomer think, *Hey, I know my ancestors found a planet around that yellow star ten centuries ago. Maybe that's who's been watching all those sitcoms*?

But this doesn't mean Kaltenegger thinks we should expect a visit. She explains this with a thought experiment she told me she often offers to her students. "I tell them, I have found two exoplanets. Both of them have signs of life in the atmosphere, one of them is 5,000 years older than us, and one is 5,000 years younger than us. And I have money and resources to go to one of those." Without fail, her students choose the more advanced planet to visit. But Earth, where we're maybe 200 years past the Industrial Revolution, less than a century into transmitting radio waves to the cosmos — we're going to be less advanced than almost any technological civilization that might be out there. "Earth is my favorite planet," Kaltenegger said, "but maybe we're not that interesting yet."

She still feels the imagined kinship of extraterrestrial observers, though. "When I was doing this work, I was thinking that if there's life out there, I really hope that somebody's rooting for us." Maybe, she mused, someone saw the Earth two billion years ago, when oxygen started building up in our atmosphere, and they kept an eye on us. More recently they would have seen us begin to destroy the ozone layer, and then fix it, and then begin to wreak havoc with climate change. She imagined them cheering us on. "*Come on, come on, fix it!* Hopefully there's some nice thoughts out there."

Not every study that looks at Earth as an exoplanet so explicitly situates us among a galactic community, but that subtext is always there. And that's part of what makes these studies so compelling. They're scientifically important, but they also presume an inhabited cosmos, taking for granted that there could be other observers out there, other beings, other people, as curious about us as we are about them.

In my conversations with researchers doing this work, often they'd talk about seeing Earth not as an exoplanet but as a planet. It could be a syllable-sparing shorthand, or scientific precision, or something more meaningful. We don't see Earth as a planet, after all, in daily life. It took math and insight for ancient thinkers to realize the Earth was a sphere and not a flat surface; it took Copernicus's intuition and Galileo's telescopes to realize that the planets and Earth alike were spheres, all in their parallel orbits. To study planets is to see Earth as one of many, among comrades in the solar system and kin in the cosmos, just one of what turn out to be bountiful worlds.

CHAPTER 3

Animals

I first read Carl Sagan's *Contact* and *Cosmos* in high school, when I was working at a bookstore that let us borrow any book we had at least two copies of on the shelves. I loved them then and was excited to revisit these books in the course of my research for this book. A scene from *Cosmos* had stayed with me — and confounded me — for almost twenty years, and I was ready to make new sense of it. But when I got to the place in the book where the scene should have been, the chapter just ended.

Luckily, I was reading an e-book, so I could search *anus* . . . Zero results.

Let me explain.

The scene I remembered is in a classroom — I don't know if Sagan was teacher or student — and the class is split into groups, each tasked with designing an alien life-form. The professor looks at their anatomy sketches and chides, "Where's its anus?"

What had haunted me across the decades was my confusion: Was the professor closed-minded, or were the students being careless? There was meant to be a lesson there, but unable to parse it I was stuck in that very teenage feeling of knowing the adult world had a meaning I couldn't access.

But now, researching an entire book on aliens, I was ready to finally figure it out. And it just wasn't there. It was even harder to google:

"Where's its anus"

"alien anus"

"anus Sagan"

Nothing. The last one was weirdly thwarted by the fact that Sagan once sued Apple for calling him a butt-head.

I thought I was beat, but I once again opened Google Books and searched "where's the anus." And there it was, with one word of the search term changed: *Sphere*. A book I'd read even before *Cosmos,* a book that's sort of about how unknowable alien life could be but is also a fun spooky thriller. *Sphere* is Michael Crichton at his (according to my seventh-grade memories) absolute best. And that pulpy paperback, not Sagan's opus, was where this formative anus story was found.

Now that I can see the passage, it's clear that the professor was meant to be at fault.

"But many animals on Earth have no anus. There are all kinds of excretory mechanisms that don't require a special orifice . . . Who knows what we'll find?" That's obvious enough that I must have understood it even when I was twelve — but it had gotten lost in my memory, tangled up in the strangeness of aliens and incomplete imaginings, things even the grown-ups didn't understand.

Sagan agreed that alien life was probably incomprehensible. He wrote in *Cosmos* that even if extraterrestrial life shared our biochemistry, it would have no reason to look at all like life on Earth. But — twist his arm! — he deigned to conjure an example, picking a planetary environment that seemed uninhabitable to more conservative minds, a gas giant like Jupiter. That planet has no solid surface, just a dense atmosphere of hydrogen, helium, methane, and ammonia, where "organic molecules may be falling from the skies like manna from heaven." Sagan proposed life-forms he called *floaters,* organic balloons, whale-sized or larger, pumping themselves full of hydrogen or hot gas to rise or sink as needed.[23] "We imagine

23 While the biology and ecology of floaters is plausible, the emergence of life on a gas giant

them arranged in great lazy herds as far as the eye can see."

Sagan was a fabulist, if not as a scientist, then as a steward of our imaginations. He wanted us to imagine weird aliens, to shatter our anthropocentric habits. If not for scientific reasons, then for spiritual ones. There's a cosmic humility to be found in understanding that we're just one of life's infinitely diverse expressions. Even if we can't imagine truly strange, truly different life, we push against the inherent xenophobia of our imaginations when we try, while what we know pulls us back like gravity.

You can read almost every alien story as being about this tension — between xenophobia and empathy, between anthropocentrism and the desire to conceive of something else. But there's also a tension between two functions of science fiction: one, imagining life that's truly alien; and two, conjuring a kind of alien life that serves a purpose for the humans involved, be it the characters or the readers.

---

could be truly impossible, alas. While mature floaters could . . . float . . . their evolution would be stymied by the lack of a surface. The convection of the planet's atmosphere would bring them through too many disparate regimes of temperature and pressure.

Before we can think about the alien characters humans might meet — the intelligent ones, if not humans, then people — we need to look at the broader ecosystems, the alien organisms imagined or hypothesized to inhabit, in all their diversity, possible alien worlds.

This is where we leave behind the solid footing of scientific research for something more like speculation. We still have science to guide us — extrapolating from what we know about evolution and biology on Earth — but we're working by analogy and tethered by a pretty thin rope.

Researching multicellular alien life isn't very practical for scientists. Instead, it's the possibility of microbial life that rules the lab. When we send probes to other planets, scientists need to have an idea of the possible forms of microbial life and their chemical signatures in order to detect them. Hemmed in by the physics of atomic bonds, researchers can reasonably speculate. The same parameters also inform how astronomers point their increasingly sensitive telescopes at planets beyond our solar system, also in the hope of detecting life's chemical traces. But it's harder to justify lab time and funding for what is ultimately an unexacting exploration of the possible forms of complex life. Scientists work on imagining alien chemistries because we have a plausible need for that knowledge. It just doesn't make the same sense to spend time thinking about what alien animals might be like.

But that doesn't have to stop you and me.

Not Wrong, but Not Real

If you'd never seen a deep-sea anglerfish, would you think it was a friendly fellow Earthling? Or would its grotesque visage, its gaping mouth and the glowing lantern protruding from its forehead, suggest farther-flung origins? What about a platypus, with its patchworked physiology — does a duck-billed, web-footed, egg-laying mammal seem natural to you? Buried in Earth's fossil records are extinct animals that match nothing of what we know of animals today, precisely because their lineages dead-ended. If you met an Anomalocaris, a six-foot-long invertebrate cross between a trilobite and a flying carpet with a pair of segmented, spiky trunks, would you think, *Ah, yes, the bountiful diversity of my home planet, Mother Earth*?

In an office on the fifth floor of the American Museum of Natural History in New York — a warren beyond public access, where the halls are lined with collection cabinets twelve feet high — Mick Ellison has been imagining what extinct animals looked like for the past thirty years. He takes what knowledge science can give him and extrapolates, adding bones to partial fossils and flesh and skin to skeletons. He's doing his best to bring these creatures back to life.

In Ellison's work, a land-dwelling whale ancestor that lived forty-five million years ago, known only from a single fossilized skull, becomes a hunchbacked creature like an overgrown boar, with a spindly tail, round belly, long snout, and a bristle of hair tracing its spine. Sinornithosaurus, a feathered dinosaur with a chicken-sized body, whose fossils show sharp, delicate bones twisted at unnatural angles, becomes on Ellison's canvas soft, almost fuzzy, and decidedly bird-like, with a posture that puts its butt high in the air and one long leg slightly raised off the ground. Combined with the animal's curious gaze, the effect is that of a gentle creature considering whether or not to approach.

Ellison isn't imagining these animals out of thin air — even when there isn't a single complete skeleton, let alone impressions left by muscle, feather, or skin. He told me, "It's all rooted in anatomy and natural science. You see relationships, not only between extinct animals and living animals today, but you can follow lineages. Even if you have

Mick Ellison

a fossil that's one-of-a-kind, and you don't have any others like it to compare it to, you can always find things that are closely related to it and then branch out from there."

The museum's goals are educational, but they also wrap information in awe. You can read every label in the hall — or you can just look at the T. rex skeleton and go *Wow*. What you're absorbing isn't the facts about a certain extinct animal; facts about what it looked like when it was alive don't even exist, anyway. Instead, you're learning a way of engaging with a more connected worlc across distance and time.

I asked Ellison what he thinks of as his goal when he's drawing an animal he's never seen. He said he likes to think of his work as factual and based on evidence, but what guides him isn't just accuracy. Instead, he said, "My main goal is to try to create something that the viewer can accept as a living, breathing animal." Especially when it comes to dinosaurs, he said, illustrations often look like dragons or monsters. The problem isn't that those are wrong, but that they aren't real.

You don't need to be a zoologist or farmer to be an expert. We all spend our lives surrounded by animals, in images and in daily life. Cats and dogs, pigeons and songbirds, lions and tigers and bears. But the animals we're most attuned to are humans. That's why we're so sensitive to facial expressions, why CGI people are so difficult to render well, and why, Ellison told me, portraiture is the most challenging form of realistic art. We have an innate sense for what's human and what's — sometimes creepily — not. Ellison said we have that for animals, too. Beyond the question of scientific plausibility — whether the bony protuberances on an animal's skull are being appropriately used as anchors for powerful jaw muscles — even nonscientists have an intuitive sense of animal realism. Even for an animal we've never seen, that no human ever saw. Ellison said, "If you slip up in just one spot, maybe with a certain detail of the anatomy — ears, eyes, posture, or something like that — the illusion just shatters." The viewer wouldn't be able to put their finger on what was off, but you'd end up in the uncanny valley of animal realism, with something that looks stuffed or fantastical. Ellison's job, then, is to be aware of "all the subtle subconscious things that make up something that we believe is natural."

Alien animals may challenge our intuition for what's natural like nothing on Earth ever could. It's an open question whether life on another planet could even be categorized as *animal* at all.[24] In science fiction, alien life is often at least a little familiar, and you can justify this with science or with narrative needs. We imagine alien animals we'd be able to comprehend, but we hope they might expand our understanding in the process.

24 Speaking in terms of physical characteristics, even, not common ancestry (*evolutionary grade,* rather than *clade,* for the biologists out there).

■■■■

The blue-and-green monkey is only on-screen for a moment. It's not even a monkey — that's a classification you, the human viewer, make by association. It's tree-dwelling, and it swings between branches counterbalanced by a long tail. Its face is familiar, with big eyes and a nose and a downturned mouth all where you'd expect to see them. It seems intelligent, too, cocking its head and fixing the camera with an indignant stare before swinging off with the rest of its troupe. But as it does, it grips the branches above itself with two pairs of hands.

Prolemuris, as they're known, lives in the forests of Pandora, the setting of the movie *Avatar*. Theirs is a blink-and-you'll-miss-it performance, but they're a crucial piece of James Cameron's story. Not the story about militant humans and pacifist aliens and the one man who can save them all, but the backdrop so ecologically rich it demands serious consideration, perhaps promotion from backdrop status. Through it, Cameron tells an entirely different story — an evolutionary one.

For the most part, Pandora is inhabited by six-legged animals, hexapods instead of the four-legged ones (tetrapods) we know

. Earth. (Most of our six-legged animals re insects.) On Earth, tetrapods dominate in size if not sheer numbers. But on Pandora, six-legged animals run the show: the iridescent direhorse, with a ridge of blue skin where a horse's mane would be; the viperwolf, a glossy black and vicious predator; the dopey tapirus, a knee-high grazer with globby ridges on its snout. All of these animals — just about all of Pandora's animals that we see — have six legs. Except for the Na'vi.

The intelligent humanoids of Pandora are strikingly . . . humanoid. Sure, they're blue and nine feet tall, but otherwise they look like muscular humans stretched out extra lithe. They have two eyes, a nose, and a mouth just where ours are, which is nothing to take for granted. Most of *Avatar*'s animals have two pairs of eyes, each set moving independently of the other. They have nostrils on their faces, but also lines of opercula — basically air-gills — on their chests for more robust breathing. But not so the Na'vi, Pandora's people. Whether for the sake of viewer empathy or motion-capture ease, the Na'vi have far more qualities in common with humans than any of their planetmates share with other Earthly life.

This is why the two seconds of prolemuris

we get matter so much. Like the Na'vi, pr
lemuris has two eyes, not four, and no ope
cula. Most importantly, though: their arms.
Not two, like the Na'vi. Not six limbs, like other Pandoran creatures. Prolemuris has . . . two and a half arms? It's hard to figure out the right math, but basically prolemuris has legs, with two sets of arms that each branch into two hands. What it looks like, strikingly so, is the partial fusion of a pair of limbs on each side. Prolemuris gives us an evolutionary link between the hexapod animals and tetrapod people, a scientific logic for what was probably an aesthetic choice. Or at least an attempt.

Speaking of attempts . . .

rolemuris's partially fused limbs are obviously a gesture at evolutionary logic, a way to get from animals with six limbs to people with four. Plenty of Earth animals, like snakes and dolphins, have arms and legs in their ancestry but none today. But all of those animals lost their limbs because of evolutionary pressure. Snakes' ancestors lived in burrows and did better with stubbier and stubbier legs until they were gone. A dolphin progenitor thrived in its new aquatic environment as it lost its hind legs, too, while its front limbs worked better as flippers. Just as human embryos sprout tails and then lose them, in a developmental recreation of evolution, dolphin embryos have, for a few moments, hind legs.

But paleontologist, author, and science educator Katie Slivensky points out that a tree-climbing monkey-type is the very last animal who would find four limbs evolutionarily advantageous over six. More hands on unfused arms would make a monkey better at climbing and swinging, faster and more dexterous. Making matters worse, Slivensky wrote in a post on her blog, "If any animal on Pandora should fuse limbs . . . it would be these ground-dwelling ones." Animals that run tend to evolve toward less contact with the ground — a horse's hoof, for example, is

actually a single toe. "But in every instan
of a ground-dwelling critter on Pandora,
of course has six limbs."

This isn't about policing pop culture for scientific fidelity, of course. Slivensky reminded me when we spoke, "A lot of pop culture goes more for the imagination than the accuracy." But, she added, if you're going to not only invent an alien creature but also depict its evolutionary context, ". . . it's actually really easy to just talk to someone for five minutes and get an accurate scientific take on something. It gives you that extra layer of realism." Even for a lay audience, scientific accuracy is one way to trigger that subconscious click, that sense that this is something real and alive.

Instead, *Avatar*'s logic seems to me to run on a spectrum from *like people* to *not*. The Na'vi are most humanoid, the common animals are less, and the monkey-creatures get halfway. But evolution doesn't follow that anthropocentric logic. It doesn't follow most of the logic human minds want to impose on it at all.

Ways to Make a Living in the World

Evolution is the driving philosophy of many fictional alien worlds, whether authors intend it or not. And when a storyteller invents an

.en world that looks familiar — with pho-
osynthesizing plants and roaming animals, similar or analogous to the kinds of animals we know on Earth — it's grounded in the scientific principle of convergent evolution.

Evolution gives us the mind-boggling diversity of life, but it also gives us a lot of animals that look pretty similar to one another, too. Unrelated species often evolve the same features independently, because those features are evolutionarily useful. Not every Earth animal with wings or a lensed eye evolved from a common winged or eyed ancestor; these animals stumbled into the same biological solutions to problems like flight and vision. Whales evolved fishlike features but are no more closely related to fish than cows are. This is convergent evolution — distantly related organisms converging on the same traits.

Would this happen on another planet, too? Not just a convergence of that planet's common forms but on forms that are common to all planets? Let's assume (big assumption) this other planet has land and water like we do, as well as maybe even plants and animals (though, even those categories would be a dramatic instance of convergence, hardly to be taken for granted). It seems like it makes sense that any animals that live in the sea would be roughly fish-shaped — it's a ve good body for swimming. On a planet wit a transparent atmosphere, light offers a fantastic way to perceive your environment, so these aliens would likely enough see, and eyes have proved on Earth to be the best way to do that. If you have tall plants like trees, why wouldn't you have animals that swing through their branches? And wouldn't long limbs and grasping hands, whether four or six or whatever, be good for that? Convergent evolution is so visible on Earth, it seems likely enough that life elsewhere would converge on these efficient forms, too. This is where a billion and a half years of trial and error have gotten us. So the supposition that life on another planet would follow similar forms as on Earth — plants and humanoids and horses and branch-swinging monkeys — isn't just a lazy fictional gloss.

But even on Earth, the dominance of convergent evolution isn't settled science. In his 1989 book *Wonderful Life: The Burgess Shale and the Nature of History,* paleontologist Stephen Jay Gould laid out the now-canonical argument against the predictability or repeatability of evolution. He writes, "Wind back the tape to the early days . . . [and] let it play again from an identical starting point, and the chance becomes vanishingly small at anything like human intelligence would race the replay."

He shows how the narrative of evolution only comes to us through hindsight. We talk too easily about *primitive forms* and *sophisticated animals,* imposing a hierarchy on a process that isn't progressive at all. Life changes through evolution, but it doesn't advance and certainly not toward perfected forms or anything close to it. The way we illustrate evolutionary relationships only exacerbates this misconception. Linking living species through their relationships, the tree of life appears to diversify and proliferate, from basic to advanced. It moves from root to branch, so: progress! But, Gould points out, the ebb and flow of extinction and explosion provides no such neat narrative.

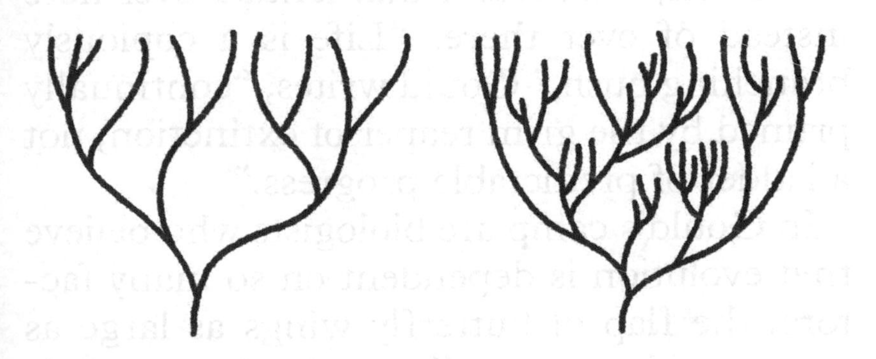

Gould's argument is driven by a reinterpretation of the fossil remains of the Burgess Shale, a rich deposit of soft-bodied creatures in British Columbia. Once thought to hc precursors of the familiar species we see o Earth today, the Burgess fauna were later revealed to be representatives of strange and — most importantly — extinct groups. They didn't evolve into us and our contemporaries; instead these diverse early organisms were nature trying out a ton of different options, and it seems like luck that the one that led to us survived. Gould argues that "we cannot use mere survival as evidence for superiority," as there's no proof that our ancestors outcompeted their contemporaries. They might've just lucked out. Any one of those other bizarre body plans could have survived to be the blueprint for eons of subsequent creatures if a predator had swum left instead of right on one ancient afternoon, if a meteor had landed over here instead of over there. "Life is a copiously branching bush," Gould writes, "continually pruned by the grim reaper of extinction, not a ladder of predictable progress."

In Gould's camp are biologists who believe that evolution is dependent on so many factors, the flap of butterfly wings as large as an asteroid or as small as a wonky molecule of DNA, that any tiny variation would have sent life in a completely different direction. Evolution, they say, is neither repeatable nor dictable; it is pure, random luck that life ere looks as it does. On the other side of the debate are those who are convinced by convergence, who believe it's an underlying rule of evolution rather than a cool thing to notice about some similar-looking species. Those arguing for convergence think evolution is predictable, that, as one evolutionary biologist puts it, "there are only so many ways to make a living in the world." These scientists believe that the same forms, proving themselves advantageous, evolve again and again, the best solutions to the challenges nature provides.

This thinking isn't limited to Earth. Some scientists believe, contrary to Carl Sagan's admonitions, that if we ever encounter alien life, we'll find it looks familiar. Biologist Robert Bieri: "They won't be spheres, pyramids, cubes, or pancakes. In all probability, they will look an awful lot like us." Astrobiologist David Grinspoon agrees: "When [aliens] do finally land on the White House lawn, whatever walks or slithers down the gangplank may look strangely familiar." And evolutionary biologist Simon Conway Morris said, "The constraints of evolution and the ubiquity of convergence make the emergence of something like ourselves a near-inevitability."

Here on Earth we do have a test case t. comes pretty close to another planet: Au. tralia. Marsupials split off from the placenta. mammals of the rest of the world about 125 million years ago, and Australia has been an island for the last 35 million, cooking up its own species in near isolation. And yet, strange as Australian mammals are when compared to their distant cousins, they also look strikingly familiar.

A common way to illustrate convergent evolution is a side-by-side comparison of a handful of Australian animals with their off-island counterparts. Australia has sugar gliders, which are basically marsupial flying squirrels — similar in the niche they fill and the anatomy they've evolved to do so. There are marsupial anteaters, mice, and moles, all extremely distantly related to the placental animals they nonetheless have evolved to so closely resemble. There's the carnivorous thylacine — or there was, until it was driven extinct — lupine in niche and appearance; and the spotted cuscus, a fuzzy, shy, big-eyed tree-dweller with a long and dexterous tail, looking for all the world like a lemur with just a bit more snout.

There's obviously convergence in these examples — the Australian animals and their doppelgängers are extremely distant cousins but I wonder if we're also seeing familiar animals because that's what we (the non-Australians among us) already know, the classificational equivalent of seeing shapes in the clouds. Australia's quoll is called a *tiger cat* but, I'm sorry, it looks more like a tiny lemur-bear-rat than it does any feline. And just as easily as you can point to the convergences of Australia, you can point in the other direction to South America, which until a few million years ago was an island as well, with no placental carnivores on its shores. There were a few marsupials, but the dominant carnivores ended up being giant, flightless birds. "We must conclude, I think," Gould writes, "that South America does represent a legitimate replay — round two for the birds." A tape rewound, and a new outcome.

But if you zoom out and squint, the sketch of convergence is still there. Some animals evolve small to live in the underbrush, while others grow fangs and claws to catch them. Maybe these niches aren't universal — after all, Australia and South America are still on Earth, they both have plants and dirt and insects, and even strange marsupials are still carbon-based, have DNA, and share common ancestors with the rest of us, however many millions of years back.

Just as carbon and water may be universally ideal foundations for biochemistry, the morphology of life may have its own optimal forms. Even Gould concedes, "Much about the basic form of multicellular organisms must be constrained by rules of construction and good design. The laws of surfaces and volumes, first recognized by Galileo, require that large organisms evolve different shapes from smaller relatives in order to maintain the same relative surface area. Similarly, bilateral symmetry can be expected in mobile organisms built by cellular division." Convergence may be just a matter of scale.

Evolutionary biologist Mohamed Noor pointed out to me that convergence is also a matter of *surface*. "When you say *convergence,* it doesn't mean things are exactly the same. It just means that with respect to the environment, there are similar-looking adaptations." So the parts of an animal that come in contact with the environment — for sharks and dolphins, their fins and skin — look similar, sure. "But if you dissect them," Noor said, "you can very quickly see it's just that piece." A shark has gills and a cartilage skeleton; even just inside its familiar-looking fins is a spray of cartilage spikes entirely unlike a dolphin's erstwhile foot bones. And a olphin's organs are quite unlike a shark's and very much like ours, lungs and larynx and twisty stretch of intestines. At least until you get to its echolocation abilities, but that's a whole other thing.

So if alien animals evolved convergently with life on Earth, we could see similar adaptations in body shape, limbs, sensory organs. It calls to mind the fiction trope of a humanoid alien bleeding alarmingly colored blood — Vulcans bleed green and Klingons purple — revealing with only a cut that they're far less familiar than they seem.

One way to imagine alternative paths for life to take, according to Simon Conway Morris, now a leading voice for convergence, is "to assess the diversity of life and ask whether there is anything one might reasonably expect to see, but seemingly has failed to evolve." He cites paleontologist and natural historian R. A. Fortey, who identified one potentially empty niche, the high-atmosphere currents "which transport insects and spiders, like some plankton of the ether." Fortey wonders if some "aerial whale" could have evolved to fill that niche, a version of Sagan's floaters, those filter-feeding predators of the skies. Many fish have gas-filled swim bladders that mediate their buoyancy — why not a similar

organ for even lighter creatures in the ar
Sure enough, in her novel *Semiosis,* Su
Burke adds to a very Earthlike alien ecosystem a suite of plantlike organisms buoyed with little sacs of hydrogen. And Conway Morris points out that hydrogen is abundant on Earth, just coupled with oxygen in water, and some microbes generate it as a byproduct of their nitrogen fixation.

But it turns out that Earth has failed to evolve its own floaters for plenty of practical reasons. Assuming the early ancestors of these sky-whales would be small (life starts small because smaller organisms require less energy) their volume-to-surface-area ratio could make them too heavy to get off the ground. And at least on Earth, the "plankton of the ether" is just too sparse to sustain a population of predators. So, Conway Morris writes, "the nearest approach to terrestrial suspension feeders are the web-spinning spiders." Sky predators, check.

Other proposed empty niches are behavioral rather than environmental. Eusociality, the colony social system of ants and bees, where everyone attends to and/or mates with one reproducing female, emerged convergently for many insects as well as coral-reef shrimp. In the mid-1970s, zoologist R. D. Alexander imagined what a eusocial ımmal might be like. From the preface of *ɿe Biology of the Naked Mole-Rat:*

> In an effort to explain why vertebrates had apparently not evolved eusociality, he hypothesized a fictitious mammal that, if it existed, would be eusocial. This hypothetical creature had certain features that patterned its social evolution after that of termites (e.g., the potential for heroic acts that assisted collateral relatives, the existence of an ultrasafe but expansible nest, and an ample supply of food requiring minimal risk to obtain it). Alexander hypothesized that this mythical beast would probably be a completely subterranean rodent that fed on large tubers and lived in burrows inaccessible to most but not all predators, in a xeric [dry] tropical region with heavy clay soil.

By the 1980s, though, eusociality had been discovered in a mammal, the naked mole rat, which fits Alexander's predictions basically to a tee.

But the most popular of these empty-niche thought experiments, Conway Morris writes, is wheels. Wheels are amazing, reducing friction and increasing leverage to push, pull, or propel a load. But no creature on Earth has natural roller skates. There are wheellike structures in biology: the rotary base of a bacterial flagellum, shrimp that curl their entire body into a loop for a few dozen consecutive rolls. But no one uses wheels to get around.

This is because, Conway Morris writes (paraphrasing biologist Michael LaBarbera), "any animal wheels of reasonable size would presuppose effectively flat and continuous surfaces," quite different from the undulating imperfections of nature. Natural surfaces are craggy, soft, and sticky. Wheels haven't evolved because paved roads haven't evolved, either. At least, not on Earth.

When Mary Malone finds herself on an alien world, she sees, striping through the grassland, "what looked like rivers of rock with a light gray surface." Upon closer inspection, she thinks, "It might once have been a lava-flow . . . It was as smooth as a stretch of well-laid road in Mary's own world, and certainly easier to walk on than the grass." Easier to walk on — and easier to roll on, too.

In Philip Pullman's *The Amber Spyglass,* Mary has slipped through a tear in the wall between worlds, where she finds an alien ecosystem. And since this world does have

, own natural roads, sure enough someone's evolved to use them.

At first, Mary is baffled by the animals she sees; in the distance, clearly, quadruped grazers with their heads down in the grass that grows between the lava-flow roads. But something about the way they move doesn't make sense. As she gets closer, she realizes it's because their legs make a diamond shape, "two in the center, one at the front, and one under the tail, so that the animals moved with a curious rocking motion."

But the true strangeness comes when she meets this world's intelligent inhabitants. They're not bipeds like humans but quadrupeds on all fours, and they have the same diamond-shaped frame as the grazers. And instead of a rocking canter, they move — though Mary tells herself this is impossible — on wheels. Specifically, wheels on their front and rear legs, while they propel themselves like toddlers on a balance bike with the two legs at their sides.

"But wheels did not exist in nature, her mind insisted; they couldn't; you needed an axle with a bearing that was completely separate from the rotating part, it couldn't happen, it was impossible —"

Well, if the world could offer up roads for

them, why not separate wheels? The mulefa, as Mary and we will learn they are called, live near groves of massive, massive trees, which drop smooth, round seedpods about the width of a human hand. "Perfectly round, immensely hard and light — they couldn't have been designed better." (The question of design — and God, and all that — is quite *a thing* in these books, but I think Mary means *They couldn't have been designed better if they had been designed, which they weren't, because this is all about evolution.*) This world gave its creatures everything they needed to evolve wheeled motion: smooth ribbons of road, sturdy wheels in seedpods, which aside from their strength and smoothness have a central hollow into which the mulefa hook a little claw; the pods secrete an oil that offers natural lubrication.

When Mary realizes what she's seeing, "she couldn't help laughing out loud with a little cough of delight." I feel the same way reading about this world, to be honest — designed, as it is, by its author, but with the satisfying click of logic we can find in the world if we see life not just how it is but as the result of eons of accidents and coincidences. The systems and beauty in their complexity and balance.

In the decades since Gould was writing, the study of evolution has become far more robust — and, indeed, experimental. No, we can't rewind the tape. But for isolated populations, scientists have figured out how to run several tapes in parallel, and even, with the help of vials and deep freezers and microbes in labs, do the closest we can to a rewind.

Evolutionary biologist Jonathan B. Losos explores these possibilities in his book *Improbable Destinies: Fate, Chance, and the Future of Evolution*. He writes, "By conducting experiments, we can see how repeatable and predictable evolution is: If you start at the same point, will you always end up with the same outcome? And if you start at different points, but select in the same way," that is, exert the same environmental pressures, "will you converge on the same result?"

Losos describes over a dozen experiments, in the field and in laboratories, that have attempted to answer this question. They involve everything from observational studies of wildlife to subjecting microbes or fruit flies to selective pressures (like depriving them of food or getting them drunk). The results are fascinating — and inconclusive. Sometimes evolution repeats itself.

Sometimes it strikingly does not. But what if evolution as we know it isn't how life on another planet develops at all?

In the conclusion of his book, Losos writes, of the possibilities of biology on other worlds, "Even if life were carbon-based and the genetic code were based on something like DNA, the rules of inheritance and evolution might be very different."

That stopped me in my reading tracks. I'd always assumed evolution by natural selection was not just a descriptive principle of life on Earth but a fundamental law of life. It's in NASA's working definition, after all! How else could life develop and change? What other forces could possibly drive it? Losos had just shattered my understanding of evolution. So I asked him about it.

He told me that natural selection only drives evolution when a system meets three criteria: there is variation within a population or species; the variations affect fitness so that some variants survive or reproduce better than others; and there is an element of genetic heritability, so that variations are passed on to the next generation. While all of these characteristics seem essential to life on Earth, there's no reason life on another planet would have to be the same.

Darwin didn't invent the idea of evolution,

he just figured out the mechanism of natural selection (and then managed to write about it more prominently than anyone else who'd had similar thoughts). A century before him, Jean-Baptiste Lamarck had proposed that as individuals changed their physiques in life, those changes would be passed on to their offspring. The classic textbook illustration, which oversimplifies Lamarck's ideas but gets the point across, is a short-necked giraffe ancestor: he stretches his neck to reach high-up leaves, and his sons have slightly longer necks, making them better leaf-eaters. Or imagine a bodybuilder. She spends her life making her muscles bigger, changing her body. Why can't those physical changes be passed on to her heirs?

For starters, Losos said, because most animals on Earth don't pass genes to their offspring from cells in their necks or arms. "The genes we pass on to the next generation come from a small number of cells that are segregated from the rest of our bodies — the ones in our gonads, which become eggs and sperm. But there's in theory no reason why other life-forms would have to restrict those cells."

We don't know a mechanism by which changing something about your body could alter your genes, but that may not even have to happen for something like Lamarckian evolution to work. Even for life on Earth, the environment and our experiences can change how our genes are expressed — basically, what's switched on or off, or which genes are being actively read for instructions. Research has shown that stress, trauma, disease, and diet can change gene expression in humans. Losos said, "There are beginning to be some indications that these epigenetic changes can get passed on to the next generation. But most evolutionary biologists think that although this may occur, it's not going to radically change our understanding of evolution."

But that's just here on Earth. He said that in an alien system, epigenetic changes could be heritable and significant enough to drive evolution. Pump your muscles, raise swole offspring.

I felt the thrill of invention — and validation, since Losos wasn't reining in my fancy but instead giving it his most generous stamp of approval: *There's no reason why not*. Even still, all these possibilities we drew from were familiar from Earth. But it seemed a rich palette to work with. Horizontal gene transfer was my next idea, that multicellular animals might somehow retain this ability known mostly in bacteria on Earth, to share genes peer to peer, rather than just parent to child through inheritance. Bacteria can do this easily because their genes are not sequestered in a nucleus, which the cells of all multicellular animals on Earth have. But could multicellularity exist without a nucleus? Perhaps the universe could see a way around these problems better than I, or even an evolutionary biologist, can. Maybe elsewhere, larger organisms like animals could be more open to errant genes. There could be a system of something like viruses transmitting genetic code willy-nilly. You could get genetic variation not just from your parents but from anyone who sneezed on you or brushed arms with you on the street.

That, Losos said, leads to the idea of an entire way of life without species. What if there were no hard boundaries between one animal and another, just "a smear of variation"? Species are taxonomical categories, but their boundaries traditionally correspond to biological lines, across which two organisms can't reproduce — meaning, for our concerns here, they can't combine their genes into offspring. But without species boundaries preventing genetic exchange, you'd have a very different range of evolutionary trajectories. "It would certainly make evolution, as a process, very different," Losos mused.

"Would that just be an impediment and make it slower, or might it cause different sorts of evolutionary patterns? I think probably the latter."

Mohamed Noor told me that a speciesless planet would actually be *more* subject to natural selection than life on Earth. "Imagine," he said, "that something is advantageous. And it spreads through the population of ants." On Earth, that new adaptation has nothing to offer humans. But a speciesless world is essentially a world without those boundaries — a world all of ants. Not identical, but not separate species, either. "Then it would spread to everything, right?" And the whole planetary population would evolve. In this case, species aren't outcompeting each other but rather genes are, the advantageous ones dispersing across the whole world.

Losos showed me that any vision of alien biology that converges at all with Earth's presumes a common mechanism of evolution. And that is hardly a given. So any alien evolution we see in fiction is more like an alternate history of life on Earth. Rewind the tape, give the primordial ooze a small nudge, and see how things progress: a six-legged vertebrate ancestor may outcompete a progenitor with four. Try as we might to imagine alien animals, what we're really doing is finding another way to understand life here on Earth.

We know a bit about Gould's and Conway Morris's personal beliefs, just enough to wonder if their understanding of science shaped their worldviews or vice versa. Conway Morris doesn't dig into this much in his writing, but it has been pointed out to me that as he is Catholic, the importance and inevitability of humanity may hold for him extra weight — there is ample science supporting his sense of things, but that sense may also be what he needs to be true. Gould, on the other hand, who for whatever it's worth was a secular Jew, writes that accepting the primacy of contingency "fills us with a new kind of amazement (also a frisson for the improbability of the event) at the fact that humans ever evolved at all." We don't need the facts of the world to be any particular way for them to be important. Humans are very good at making meaning of things.

The aliens we imagine in fiction are just as much vessels for meaning. It's not that storytellers try to make something we'd recognize, but that recognizable alien animals serve a narrative purpose — they appear welcoming both to human characters and to the human audience. When the familiar is graced with a flourish of strange, it reads as intriguing or whimsical more than anything else — a reminder of the grand diversity of the cosmos (and, by extension or implication, of Earth). But not every alien world is meant to be welcoming. And not every alien animal is meant to satisfy our innate sense of what's natural. Sometimes, of course, it's the opposite.

The Most Alien Aliens

When I think about most of my favorite stories set on alien worlds, I realize the animals are part of the background. Even prolemuris can only be investigated in freeze-frame, its time on-screen is so brief. Most of the time, animals aren't characters in science fiction. They're part of the environment and part of what makes a world feel strange or familiar to us.

One very silly version of an alien animal exemplifies the path to strangeness. The Alfa 177 canine, from a 1966 episode of *Star Trek,* is a scruffy blond terrier outfitted with a lion wig, antennae, and a unicorn's horn. His whole look says *Fine, is this alien enough for ya?* Whether he is evidence of limited imagination or limited budget, a vague production-design shrug or a joke, this canine embodies one of the most powerful ways of making animals seem alien:

blur the categories, break the rules. In this case, he's clearly a dog in a cheap Halloween costume. But his noncanine qualities aren't random, they're deliberately drawn from other animals, some only distantly related to dogs. He's not just the alien equivalent of an Earth animal, a blue monkey or a horse with an extra set of legs. He's something else entirely.

Blurring the Earthly categories we know is a surefire way to trip the alarms of our intuition for naturalness. But Losos suggests that this may be exactly the kind of convergence we're most likely to see on another planet. In his book he uses the example of the duck-billed platypus, in many ways one of Earth's strangest one-offs. An egg-laying mammal, endowed with venomous spikes in its armpits and a sixth sense that lets it find underwater prey by the wafting electrical signals of the nervous system. Very weird! But at the same time, familiar in its patchworking: a beaver's sleek body, a duck's paddling feet, the titular bill. And the platypus isn't the only animal to match bits and bobs of its anatomy to others through convergence. Human and octopus eyes appear nearly identical, set into wildly different bodies. "Here on Earth," Losos writes, "species frequently do evolve similar features in response to similar environmental conditions." So while he thinks humanoid — or platypusoid, for that matter — aliens are hardly a sure bet, "that's not to say that extraterrestrials would look completely unfamiliar. An extraterrestrial might even be a mash-up, platypus-style, of many different parts borrowed from different Earth inhabitants."

But while the Alfa 177 canine is adorable, other alien hybrids are decidedly not. With better special effects and better storytelling, the unnatural alien becomes plausible, too — and that's when things take a turn.

The Xenomorph from the 1979 film *Alien* is one of the most disturbing aliens to ever grace the screen. From our first shadowed glimpse of it, we know this creature is utterly unnatural, and yet it seems powerfully real. It passes Mick Ellison's naturalness test and seems somehow plausible, yet still strikes us as alarmingly wrong. Psychoanalyst Harvey Greenberg writes that this alien disturbs us because it defies "every natural law of evolution: by turns bivalve, crustacean, reptilian, and humanoid." But the laws of evolution Greenberg cites aren't laws at all — they're Earthly categorizations, a narrow definition of *natural*. There's no evolutionary rule against hybridizing the categories we know. There's no evolutionary rule against new categories either, by whatever mechanism your evolution works. It's only our Earth-bound framework that makes a bivalve-crustacean-reptilian-humanoid seem unnatural to begin with. The Xenomorph is horrific even before it tries to hurt you, as soon as you get a glimpse of its shape. It's less an animal than a monster.

Monstrous doesn't just mean scary or gross. In his essay, "Monster Culture (Seven Theses)," scholar Jeffrey Jerome Cohen lays out a taxonomy of monsters — not of their physiologies but of their roles in culture. Monsters, Cohen writes, embody and enforce a culture's anxieties and desires. For Cohen, the Xenomorph violates the taxonomic, evolutionary categories that Greenberg lists, but I think it's even more disturbing for violating another boundary that's crucial to our sense of life on Earth. The Xenomorph is intelligent, able to sabotage human technology and communicate among its kind, but it's driven by parasitic reproductive instinct and protective violence. It blurs the distinction between animal and person.

We draw lines between people and animals for environmental, economic, legal, and culinary purposes, but those lines are not always firm. And so a monster is a plausible unnatural creature, violating familiar distinctions and daring us to believe its impossible biology. Then inviting us to take what we've learned of an alien world with us when we return home.

Imagining alien life requires a leap of faith that's not just cognitive but empathetic, too; not just assembling physiology but believing in impulses and desires, perhaps even inner life. And we have enough trouble doing this on Earth, where the animals actually are our very distant cousins.

Sometimes it's easier to practice expansive inclusion on another planet. *Avatar* gets compared to the 1992 eco-fairy cartoon *FernGully* not just for its plot and rainforest setting but for its message, one of stewardship of the Earth and respect for nature. Pandora's forests are inhabited by diverse and vividly imagined creatures. They're not a backdrop but the heart of the story — they are what is being saved. Leave the theater and, the hope seems to be, perhaps we'll see our own environment is just as precious. We're made to care about this imaginary world not for its own sake but for the sake of ours.

We hope for an inhabited cosmos because the alternative feels too lonely. We seek kinship with the inhabitants of other worlds. But that presumes that they would feel like kin. We don't even know if evolution on Earth is mainly convergent or divergent, repeatable or random. The possibilities for life elsewhere could be infinitely more strange. But if we can forge a sense of kinship with an alien animal, maybe the animals here at home, human beings included, have a better shot.

The Earth is full of alien-seeming creatures: jellyfish and tardigrades and microbes that thrive in seemingly deadly conditions, highly acidic or hot or freezing, or crammed into the pores of buried rocks. A six-legged horse has nothing on them. But it's not just the alien body that entrances and repulses us, that challenges us to imagine something beyond the boundaries of what we know. There's also the alien mind.

CHAPTER 4

PEOPLE

The human colonists who land on the planet they name Pax find much about the world that they recognize. In Sue Burke's novel, *Semiosis,* this small crew has left a struggling Earth in hopes of starting fresh, and it seems they've found an ideal world, abundantly alive but with no people, a second chance at living in harmony with an alien nature.

And the Pacifists, as they call themselves, find it easy to make sense of the life they meet. Crablike creatures scuttle through the grass, lizardish animals lives in the trees. Flightless birds hop and bats soar. The colonists name most of the life they find after the Earth life it resembles in form or function — the crabs and bats and lizards, "eagles" that hunt in land-bound packs, edible wheat and colorful tulips, lentils that happen to grow on trees and orange trees that sprout ground-reaching branches in arcs from their owns. But some of the life is too strange to easily analogize. Furry green cat-sized creatures that hop like gazelles are called fippokats after a colonist's imaginary childhood creation. (When larger, kangaroolike relatives of fippokats are discovered, the colonists name them fippolions.) Spiky organisms the colonists call cacti float aloft on hydrogen-filled sacs.

And then there are the plants.

The plants, in many ways, are familiar to the humans. They are green. They grow tall in forests and grassy in fields, they flower and they fruit. But they also act in ways that seem to at least some of the colonists alarmingly intentional. Familiar, in a sense, but nothing any plant on Earth ever seemed to do.

The first clue is in the white vines whose fruit the colonists eat. One day, three colonists who were collecting fruit are found dead, poisoned by a sudden change in the fruits' composition. A month or two later, the settlers' wheat field is flattened with root rot, tainted by water flowing downhill from a stand of the same plant. Octavo, the human botanist, finds the graves of the women who'd been killed by the fruit shot through with the vines' roots. "The vines had sent out roots to feed on dead humans," he realizes, "to tap flesh for food and blc for water." It seems, to him, like "an Ear war where corpses were left to be scavenged." He chops the roots to bits in a rage and feels foolish. They're just plants. But he also knows that with Pax a billion years older than Earth, its plants have had a billion years extra to evolve past Earth's plants as well.

"We know what plants do," one colonist says. "They grow. They're useful or they're not. And that's all we need to know." But Pax's plants have wills of their own. The vines aren't at war with the humans, but Octavo realizes they're at war with one another, one stand feeding the colonists but another poisoning their rivals' new pets. For a generation, the Pacifists fertilize the friendly vines in exchange for protection. It's a simple system of mutual benefit. But then the humans find the rainbow bamboo.

When a couple of humans camp near it one night, this bamboo sprouts a rainbow shoot in seeming greeting, and offers them delicious fruit they've never seen it produce before. But when they stop eating the fruit, the humans go through what feels for all the world like withdrawal, headachy and tired. Was the fruit a ploy for dependence, the bamboo's attempt to keep the possibly elpful humans nearby? Or was it just a plant like any on Earth, no wilier than a tobacco leaf or coffee bean?

As the Pacifists begin to suspect the plant's intelligence, the reader gets the surest sign, in a point-of-view chapter. "Growth cells divide and extend, fill with sap, and mature, thus another leaf opens. Hundreds today, young leaves, tender in the Sun." The rainbow bamboo isn't just reactive or solving problems, it's self-aware. "In joy I grow leaves, branches, culms, stems, shoots, and roots of all types."

This plant is aware of the other life around it, too, and it recognizes that the new foreigners are smarter than Pax's animals — and, thus, more useful to it. "Unfamiliar in body chemistry, but decipherable. Moths brought me bits of flesh and I learned . . . Ours will be a rewarding relationship."

The Pacifists name the rainbow bamboo Stevland, the name of the first colonist to die, an honor that they were saving for Pax's most important species. (Stevland the plant is genderless — hermaphroditic, really — but uses the pronouns of his namesake.) A human name, but there is nothing humanoid about him.

The inspiration for Stevland came to Burke when she noticed that one of her houseplants had killed another, lethally wrapping a vi… around its neighbor. At first Burke felt guilt… for not protecting the victim. But a month later, it happened again: one plant sent roots into its neighbor's pot, and into the neighbor itself. And Burke realized three things: it was a pattern, it wasn't her fault, and it could be the seed of a story.

In the research that followed, Burke learned many things about plants. She learned that they are aware of their environments. Some interpretations of plant behavior suggest that plants can measure time, interact with each other, and fight to survive. She told me, "I discovered that plants are horrible. And also amazing."

And Stevland is indeed horrible and amazing. He sees the Pacifists at first as objects of domestication, creatures to whom he can offer nutrition and protection in exchange for fertilization that allows him, as an animal never could, to grow more and more intelligent. Burke told me that, going into her research, she knew that plants need iron for photosynthesis, and in studying planets she learned that not all have abundant iron on their surfaces, as Earth does; sometimes an otherwise Earthlike planet's iron may all sink into its core. So, Burke thought, "Plants need iron, and there's a lot of iron in me.

n Earth there are hundreds of carnivorous plants, but if you have intelligent plants who eat, [they'd be] looking at me as a food source."

Sure enough, Stevland's relationship with the animals around him is, well, not quite predatory, but a kind of cultivation. He muses, "Many animals need iron just as we plants do, and iron-rich animals are nutritious. History says first we killed these animals with poison, but as we grew more intelligent, we trained them to live and die at our roots as service animals, a steady though slow supply of iron."

When Stevland meets the intelligent animals that have visited his planet, he sees them as similar candidates for domestication. But because they are smarter, he can ask for more of them, and likewise get more. "Their intelligence astounded me, far above that of other animals and plants. I could not have become what I am without their irrigation, protection, excretion, and compost."

As a plant, Stevland is not beholden to the generational life cycle of animals. He is functionally immortal, awareness spread among clones and shoots, memory stored in roots that he feeds carefully to sustain his knowledge. "Intelligence wastes itself on animals and their trammeled, repetitive lives," he thinks. "They mature, reproduce, and die faster than pines, each animal equivalent to its forebearer, never smarter, never different, always reprising their ancestors, never unique."

But still, Stevland recognizes something in the human visitors — utility, yes, a new way to serve his needs, but kinship as well. And his excitement at their arrival blurs the two. He wants them to stay. He is hungry, and he is lonely. (He is also the last of his kind, the rainbow bamboo having taught their earlier "service animals" to fight their rivals with forest fires.) He sees the colors on the Pacifists' clothing and recognizes a common sense. "They see colors. They will see mine, grand and compelling, and know . . . that I have a significant and inescapable communication to enter into with them."

And so he signals to them, a rainbow that one Pacifist realizes isn't an accidental surface shimmer but a deliberate pattern. Plants can sense light to grow toward it, after all. "This one made colors on its bark to show something." A four-year-old in the colony says, "It likes us. That's what the colors mean, and we should say we like it back."

Liking is still quite a distance away, but their relationship is mutually beneficial. The Pacifists provide Stevland with water d fertilizer; he gives them fruit produced with biochemical virtuosity, customized with stimulants, painkillers, or drugs. Through pollen and root systems, he patrols the boundaries of their territory, reporting marauding eagles or wildfire.

But for many decades, it is a tense accord. Even as the Pacifists come to communicate more and more fluently with Stevland, they are wary of trusting him. He is intelligent but too alien, too unknowable.

As Stevland becomes more involved in human affairs, the leader of the Pacificts offers to him, "You could become a citizen of Pax. You could vote, and you could take part in the debates." It's not because he seems so inclined to communality, but because his sense of superiority and control is becoming troubling. "I am not an animal," he responds cryptically. But soon enough he declares his intention to join. He tells the leader, "I have examined the polysaccharide in my most active roots and come to conclusions about equality . . . Equality is not a fact, like the length of days. Clearly I am superior to you in size and age and intelligence. Equality is an idea, a belief, like beauty."

I think the same can be said of what makes a person. Stevland is intelligent before the humans come, if wracked by solitude. But

meeting the humans, becoming a Paci
himself, changes him. Partly, it's that h
learns something of human philosophy, expanding beyond his endemic ideology of
selfishness versus mutual support to adopt
alien concepts like empathy and love (as well
as their darker contrapositives). But there's
something contractual about it, too. When
Stevland is granted citizenship in Pax, he's
granted rights and is subject to laws, and
he becomes a member of a community.
Citizenship hardly eliminates conflict and
miscommunication between the humans
and the sentient, immortal, domineering
plant, but it changes everything about how
they can seek resolution. The intelligent
alien becomes a person, too — not because
he signed a charter or because he joined a
community but because he became seen as
someone with whom those sorts of relationships could be possible.

The anthropologist and nature writer Loren
Eiseley writes, "There is nothing more alone
in the universe than man." Surrounded by
animals, he says, we humans find ourselves
separated from our natural kin by self-
awareness, by language, by history. (This is
why "in science fiction [a human] dreams of
worlds with creatures whose communicative

ower is the equivalent of his own.") The fact that we are the only ones on Earth who seem to long for this kinship makes its absence even more painful.

We see personhood in the presence of a fellow human or, some people say, in the gaze of a whale or a dog. But our track record for affording humanity to even other humans is pretty spotty, as histories of race and disability and gender clearly show.

The question of what makes a person can be approached in different ways. There's legal personhood, a category that affords rights and responsibilities — in some cases, granted to nonhuman entities like corporations and rivers; in others, petitioned for on behalf of animals like elephants, orangutans, and chimps for something like dignity. There's moral personhood, a category philosopher Mark Rowlands explains as afforded to an entity "to which a certain sort of moral consideration is owed." But most essential is what Rowlands calls metaphysical personhood — not how a person should be treated but what a person, essentially, is.

A little over three hundred years ago, John Locke defined a person as "a thinking intelligent being, that has reason and reflection, and can consider itself the same thinking thing, in different times and places." So a person, by that definition, is anyone with coherent mental life. And this is what many investigations of animal intelligence seek to measure — what Rowlands calls "reflective self-awareness," the awareness of the self as "the thinking thing." That first section from Stevland's point of view, then, isn't just establishing him as a person through narrative convention, it specifically shows him as aware of his own existence. "Growth cells divide and extend, fill with sap, and mature, thus another leaf opens." And Burke makes us wait for a few lines for this: "In joy I grow . . . " The *I*.

It is impossible, unless they're a novel's narrator, to know another being's sense of its *I*. A common way to try, to evaluate self-awareness, is the mirror self-recognition test. A spot of dye is put on an animal's body in a place they can only see with a mirror. Shown themselves — and the spot — in a mirror, a self-aware animal will realize that the reflected body is theirs, and use the reflection to investigate the spot, or try to remove it. "It is generally accepted," Rowlands writes, "that humans over the age of 18–24 months, common chimpanzees, bonobos, and orangutans consistently pass the test." (Gorillas often fail, perhaps because they are so averse to looking one another in the face other than aggression.) Claims have been made, too, or elephants, dolphins, pigeons, and manta rays. Even some fish, recently, which leaves many to question the usefulness of the test entirely — or our expectations for which animals might possess self-awareness. But we must be sure not to infer too much from the mirror test. It's often cited as proof of self-awareness, of consciousness, of thought. But what it really measures is: *Can a being recognize their physical self, and do they know that what they see in a mirror is a representation of their own body?* Perhaps the fish aren't doing anything so strange.

But we still feel that impulse to gate-keep. Rowlands writes, "Personhood is typically regarded as an accolade worthy of bestowal only on the (self-appointed) crème of the animal kingdom: humans above a certain age, and possibly (although, here, one might have to hold one's nose) other hominids and cetaceans."

Rowlands thinks we should sketch a broader circle, welcoming even creatures who can't pass the mirror test into the personhood club. But for all that animals may be our equals in terms of intelligence or worth, we have clearly gotten to the point that they are not our equals in power. So another way to understand a person is as an entity with

whom you can enter into a contract — ar
I mean that not as a legal framework but
sort of ethical one. A social contract. With
a person, your choices are not just *dominate*
and *care for.* A person is an entity whom you
must treat.

NONTERRESTRIAL

Psychologist Diana Reiss, director of the Animal Behavior and Conservation program at Hunter College, has spent her career studying the intelligence of dolphins. The core tenet of Reiss's work is to approach dolphins with no imposition of expectations. (She calls them her research collaborators.) Other researchers have shown that dolphins can be trained to understand complex cues with food rewards; Reiss says she has sought to meet them face-to-face, hand-to-fin, as equals, on a social footing she hopes is conducive to communication.

That's been one of the major revelations of the last few decades: the importance of socialness as a catalyst for the evolution of intelligence. In work published in 1960, Eiseley marveled that dolphins should have evolved to be so intelligent since they have no hands with which to make tools or interact delicately with their environment. "It is difficult for us to visualize another kind lonely, almost disembodied intelligence floating in the wavering green fairyland of the sea — an intelligence possibly near or comparable to our own but without hands to build, to transmit knowledge by writing, or to alter by one hairsbreadth the planet's surface." But intelligent animals on Earth aren't just problem-solvers, they're social beings, navigating complex webs of relationships as they make their lives, on land or sea or air. And, indeed, Simon Conway Morris identifies convergence on high intelligence as tracking with complex social structures and not, as others have argued, with dexterous hands and opposable thumbs. Reiss told me, "One of the things that we found across dolphins, chimps, and parrots is that social interaction is the driver." (The octopus is the intelligent, solitary outlier.)

Reiss sees dolphins not as an alien analogue, but alien intelligence right here on Earth — not extra- but nonterrestrial. Over her career, Reiss has worked with many SETI scientists, but even before then she said she asked herself, "If I was going to come face-to-face with an alien, or face-to-tentacles . . . what would I do? And I thought, well, one of the things I want to do is give them some way of communicating back to us."

Sometimes, dolphins find that way on

their own. One of Reiss's favorite stories from her early research, when as a graduate student she was tasked with training a dolphin named Circe to stay put while Reiss fed her. The feed was a bucket of mackerels, much bigger than Circe's mouth, so Reiss cut them into thirds — heads, middles, and tails. She noticed Circe never ate the tails, so she cut off the tails' spiky fins, which made them more palatable.

"In the course of feeding Circe, if she left the area while I was still feeding her, I had to find some way of communicating you've done the wrong thing." So Reiss essentially would give her a time-out. If Circe swam away, Reiss would walk away from the edge of the tank for a minute. Once Circe returned, so would Reiss. Eventually Circe learned.

Then, one day, Reiss forgot to cut the fins off one of Circe's mackerel tails. "It was my mistake. Circe looked up at me, her eyes were wide, she spit out the fish, and she made a beeline across the pool and took a vertical position and just stared at me. Was she giving me a time-out?"

Reiss recognized that this anecdote was just that: an anecdote, not data. But a few days later, she — purposely, this time — gave Circe tail fins, and every time, Circe ve her a time-out. Reiss felt, then, that she wasn't just studying the dolphins; she realized that they were trying to understand humans, too. That's when she realized, "these are big, beautiful, intelligent animals, and we need to find new ways to study them."

She developed an underwater keyboard that allowed the dolphins to signal for a particular activity or object to play with — a ball or disk or a nice belly rub from a trainer. Reiss found that the dolphins quickly learned to imitate the keyboard's sounds and made those sounds when playing with the corresponding objects. By the second year of the study, the dolphins were combining sounds — whistling *ring-ball,* for example, when playing with, you guessed it, a ring and a ball at the same time.

Other researchers take the opposite tack and study dolphin communication by removing humans as much as possible from the picture. Jason Bruck, an expert on dolphin cognition and communication, told me that his goal is to understand how dolphins relate to one another. "If you try to communicate with them," he said, "you are the story. And I'm more interested in them as the story."

So far, we know less of that story than you might think. We typically think of dolphin communication as a rich vocabulary of squeaks and whistles, perhaps even approaching what we'd call language. But researchers have definitively identified only one kind of whistle, the signature whistle, which is a unique identifier, somewhat like a name, though used rather differently. (A dolphin mother fetching her baby whistles her own signature whistle, for example, not the baby's.) Dolphins have plenty of other complex whistles. Some seem to be related to food, others with states of emotional arousal. But, Bruck told me, "We don't know what the nonsignature whistles mean, if they mean anything at all."

Decades of painstaking research — what Bruck calls "blue collar science" — and advances in technology have led to the understanding of signature whistles as unique identifiers (which dolphins give themselves in their first year). Dolphins can mimic each other's whistles pretty effectively, though, and they can't recognize each other by voice and tone since their voices change with the depth of water — swimming deeper, the greater water pressure raises the pitch of their voice. So instead, they rely on the contour and shape of a whistle.

With that known, it was still unclear if a dolphin reacting to another's signature whistle was recognizing the whistle's familiarity or truly hearing it as a name, a whistle stand-in for an individual. But Bruck was able to use another facet of dolphin communication that he'd previously identified, one far more alien to us: dolphins can recognize each other by their urine. This gave Bruck a way to test signature whistles. In his study, dolphins were given a urine cue and then played a signature whistle, all from other familiar dolphins. Sometimes the two signals matched up, sometimes they didn't. The researchers looked to see if dolphins would react differently to matches and mismatches. And they did, paying attention and looking around when the urine and whistle matched. A reaction that indicates, as Bruck put it, "I know that call, and I know that call is assigned to this dolphin" (whose urine I have just tasted).

Bruck credits his resistance to anthropomorphizing dolphins with his ability to devise this experiment. "I can figure out the dolphins are tasting each other's pee and getting social information from that because I'm not thinking of them like people with fins." Which isn't to say that Reiss or other researchers who try to understand and communicate with dolphins are doing things wrong — the socialness that has helped dolphins evolve their intelligence can al- allow them to connect with human trainer researchers to facilitate new discoveries. Kelly Jaakkola, director of research at the Dolphin Research Center in the Florida Keys, says that dolphins' interest in humans is a stronger incentive to engage in a researcher's experiment than food rewards ever could be. In one study, her team trained dolphins to interact with a setup with food rewards and praise. "Then we gave them the machine 'by themselves.'" When the dolphins completed the task, they'd get fish from a tube instead of from a trainer. But one particularly people-motivated dolphin gave up after a few minutes. "She looked around and was like, 'I'm not doing this anymore.'"

But despite our capacity for interspecies connection, dolphins and humans are still very different beings. Jaakkola points out that humans and dolphins are separated by ninety or ninety-five million years of evolution, and these divergent paths led to divergent bodies, brains, and ways of experiencing the world. It's not just our anatomy and senses — our brains, though both large, are very different, too. Dolphin cerebral cortexes have more area than humans' but are thinner; theirs have five layers of tissue th fewer cell types, while ours have six layers and more diverse cells. These and other architectural differences create functional differences, too. Our minds just work differently.

Measuring intelligence is only dwarfed in futility by endeavoring to rank it. Jaakkola said, "It's kind of like if you asked, 'Who's more talented, Michael Jordan or Mozart?'" A dolphin can know you're pregnant just by looking at you (or rather, echolocating into you), but if you put an object in a sound-opaque bucket and move the bucket, even if the dolphin watches you do it, they won't know where the object has gone. It seems bafflingly obvious to us, and, indeed, Jaakkola and her team assumed dolphins would easily ace this object-permanence test when they ran it. But when we don't center human abilities, there's less to take for granted. "In their world," Jaakkola said, "there aren't a lot of things that go into containers that they move." Tens of millions of years in the ocean will drive many kinds of intelligence, but tracking objects in buckets isn't one of them. Who knows what other kinds of capacities they have that we can't even imagine? So whether researchers seek to communicate with dolphins or to decipher their systems, they must do so with an open mind. Reiss told me, "I go to them, I watch them, I listen to them, and I try to figure out ways that we can work together, so they can show us what they're capable of." She says that she seeks "glimpses of the dolphin mind."

Cetaceans — Reiss's *nonterrestrials* and, as philosopher Margret Grebowicz calls them, *whaliens* — have been a point of fascination for astrobiologists for decades. The first SETI meeting in the United States included in its small attendance a prominent dolphin researcher, and the party was so taken with his findings that they called themselves the Order of the Dolphin for years after. Neuroscientist and cetacean advocate Lori Marino is a common presence at SETI conferences today, offering not just Earthly analogy but often a caution against hubris: we don't even understand other intelligent species *here*.

Dolphin research shows how challenging it could be to connect with extraterrestrial people. The fact that dolphin minds are not transparent to us does not mean there's nothing there. But decades of research have not opened a channel between humans and these obviously very intelligent animals.

The idea that cetaceans and humans can't communicate, however, is a Western one. In an article for *Hakai Magazine,* Krista Langlois writes that whale-hunting Arctic ultures saw the practice "as a match between equals." Humans had their technology, but whales had size and were seen as "emotional, thoughtful, and influenced by the same social expectations that governed human communities." Whales were thought by their hunters in these communities "to live in an underwater society paralleling that above the sea." The fact that whales were hunted wasn't a sign of human dominance, but of cooperation, whales offering themselves to worthy hunters so that their bodies might be treated with respect and their souls reborn.

Reiss also says that, since ancient times, people have sensed that dolphins are "aware of what they're doing," observing us just as much as we do them. In ancient Greece, dolphins were said to save sailors who were lost at sea. Reiss says these legends suggest that dolphins are "minded."

But Western science is constrained, Marino writes, by the *scala naturae,* Aristotle once again pinioning us with anthropocentric ideas. In this case, it's the idea of hierarchical progression in the natural world, from inorganic material to plants to invertebrates to vertebrates "and, finally, humans, who occupy a separate, superior position above all other animals. Humans are considered the most *advanced* life-form on the planet, having a more *perfect* form than the other animals and possessed of unique qualities.' In astrobiology, she cautions, this narrows our imaginings because we incorrectly think we have just one example on Earth of intelligent life.

Anthropologist Kathryn Denning goes further, reminding the astrobiology community that humans have a pretty terrible track record in our dealings with Earth life, no matter how intelligent. "Human appreciation of other life-forms and their abilities/properties leads to webs of control, commodification, and exploitation," and science and technology are tangled in those webs as well. Denning samples the cruelties we've enacted on our closest kin, like sending drugged chimpanzees into space. When we attempt to communicate with dolphins or other intelligent animals, we confront the limits of our empathy alongside language and imagination.

In her short story "The Author of the Acacia Seeds," Ursula K. Le Guin writes fragments of a future scholarly journal, the work of the Therolinguistics Association. *Thero* from the ancient Greek for *feral* or *beastly* — a linguistics of wild animals, and even wilder things.

The first excerpt is a scholarly review of a manuscript written by an ant through "touch-gland exudation" on an orderly row of acacia seeds. The human scholars recount the academic dispute over the manuscript's final line: "Up with the Queen!" They propose that it must be read counter to human ethnocentricity: While *up* means exultation to humans, for ants "*down* is where security, peace, and home are to be found. *Up* is the scorching sun; the freezing night; no shelter in the beloved tunnels; exile; death." The corpse of a single worker ant has been found beside this final seed, perhaps the seditious author.

In the next excerpt, Le Guin ventures further from human language, in the voice of a scholar proposing an Antarctic expedition to study the emperor penguin. The language of the Adélie penguin has recently been deciphered thanks to high-speed cameras, which can catch the "fluid sequences of the script" of the "group kinetic performance." (Translation is still challenging, but the best attempts have been made by the full corps of a ballet troupe.) The author proposes that studying the isolated, aboveground emperor may be more fruitful: "In that black desolation a little band of poets crouches . . . Like all kinetic literatures, it is silent; unlike other kinetic literatures, it is all but immobile, ineffably subtle. The ruffling of a feather; the shifting of a wing; the touch, the faint, warm touch of the one beside you."

But even the kinetic literature, the poetry of warmth and existence, is not the strangest thing therolinguists seek. The story ends with an editorial from the president of the Therolinguistics Association, who challenges the reader to imagine past the limits of language, to art beyond communication. (The therolinguists study literature as much as they do language.) A new frontier for the field may be found there, in "the almost terrifying challenge of the Plant." (A very different kind of terror from Stevland, a terror of radical empathy and scholarly ambition.)

The president dares the reader to consider plants' art, with a reminder of past linguists' foolish provinciality. "Remember that so late as the mid-twentieth century, most scientists, and many artists, did not believe that Dolphin would ever be comprehensible to the human brain — or worth comprehending! Let another century pass, and we may seem equally laughable." And those future therolinguists will laugh indeed, "they will smile at our ignorance as they pick up their rucksacks and hike on up to read the newly deciphered lyrics of the lichen on the north face of Pikes Peak."

But even that is not a wild-enough curiosity. Beneath the "delicate, transient lyrics of the lichen," the president proposes, someday the first geolinguist will read the poetry of stone and earth itself, "in the immense solitude, the immenser community, of space."

There is always another depth to pursue, reaching farther and farther from ourselves until solitude and community converge.

Alien Forms

Alien people in science fiction tend to take familiar forms, even when they don't look like humans. Bug-people, snake-people, ape-people, lizard-people. They answer the implicit question, *What if these were the lineages that evolved intelligence on their worlds?*

In 1982, paleontologist Dale Russell proposed that if dinosaurs had not gone extinct but had remained dominant on Earth, intelligent life would have evolved in their descendants. He went so far as to sketch what he called a dinosauroid, following evolutionary logic to a strikingly humanoid form. Intelligence requires a bigger brain, a bigger brain goes along with a flatter facial region, so no snout, and a heavier head is better balanced on an upright body. Then walking upright gives you some familiar-looking legs, and your claws evolve dexterously toward fingers. Jonathan Losos describes the resulting sketch as having "an uncanny resemblance to a human, right down to the butt cheeks and fingernails."[25]

It's convergence, again, but taken a few steps further. Just as wings and eyes evolved multiple times on Earth, might personness evolve multiple times across the universe? And would humanness go along with it? Simon Conway Morris says don't worry about the dinosauroids, anyway — if the asteroid hadn't killed off the dinosaurs, the ice age would've, and humans would have just been thirty million years behind schedule.

Conway Morris's argument is right in the title of his book, *Life's Solution: Inevitable Humans in a Lonely Universe*. We, or something quite like us, were inevitable on Earth due to the forces of evolution. He identifies the likely convergences as "large brain,

25 Losos later took up the same thought experiment, with a much more avian result — with what we know now about dinosaurs as bird ancestors and deliberate resistance to the strong pull of the humanoid, his dinosauroid looks like a big-headed, wingless sparrow, clutching a spear in a scaly front claw.

intelligence, tools, and culture,"[26] which even on Earth are converged upon, evolved into again and again. (See: dolphin, chimpanzee, octopus, crow.) And he thinks that since life elsewhere would be Earthlike, convergently, then from there he feels sure humanlike life will arise on other planets, too.

We may then, actually, find intelligent life on another planet to be *more* familiar, or comprehensible, than the other intelligent life here. For all the points of connection — the interspecies relationships, the trainability, the ineffable something that humans seem to recognize in cetacean eyes — whales and dolphins may be truly alien because, as Reiss reminds us, they're nonterrestrial. If an alien evolved in an environment more like ours, even just on land instead of in water, it's reasonable to anticipate convergence.

Perhaps we're lonely on Earth because we've outcompeted — or subsumed — anyone who might've really kept us company. A hundred thousand years ago there were other species of people on Earth. Neanderthals and Denisovans were similar enough to us to mate with, as their ghosts in modern human DNA attest. But now all that's left are those species we can't speak with.

26 So, yes, a dinosauroid would qualify.

Maybe every inhabited planet has its one lonely species, looking to other worlds for kin. If there's convergence on the humanoid, there could be further convergence on a kind of intelligence, a kind of thought, a kind of language. Though whispers of our disparate origins may remain.

When Lilith, the human protagonist of Octavia Butler's novel, *Dawn,* first sees an alien, her reaction is beyond fear, beyond disgust. It is "a true xenophobia," Butler writes, the recognition that what she sees is unnatural, at least to the nature Lilith knows. The alien is humanoid, the size of a tall, slender man. He's gray, covered in what looks like hair all over his head, eyes, and ears. He makes no frightening moves, he even seeks to guide Lilith through her revulsion, gently pointing out that what she probably thinks is hair is a mass of delicate tentacles. He invites her to look, and that's when it hits her. "She did not want to be any closer to him. She had not known what held her back before. Now she was certain it was his alienness, his difference, his literal unearthliness. She found herself still unable to take even one more step toward him."

As the natural history artist Mick Ellison proposed in the last chapter, we innately know what a plausible animal is and isn't. Butler proposes that our sense of what's natural is confined to what is natural to Earth. The alien's hair turns out to be sensory tentacles, and though Lilith compares them in revulsion to night crawlers, Butler writes of a deeper wrongness resonating at the meeting of people from distant worlds. Lilith will eventually overcome her horror, but as close as she becomes with these aliens, essential differences will always remain.

Lilith has been saved, along with humanity's scant survivors, from the aftermath of a nuclear war. Their rescuers are an alien race, Oankali. One of them, named Jdayha, tells Lilith that the name of the people also means *traders* in their language.

"What do you trade?"

"Ourselves."

"You mean . . . each other? Slaves?"

"No. We've never done that."

"What, then?"

"Ourselves."

The Oankali currency, Lilith will learn, is genetic code. "We acquire new life — seek it, investigate it, manipulate it, sort it, use it," Jdahya tells her. But this is not a technological power. "We carry the drive to do this in a minuscule cell within a cell." It is an organelle in their cells that allows some members of the species to understand and manipulate DNA.

The Oankali trade with other species for the genes they need. It's a souped-up, galaxy-spanning version of horizontal gene transfer. Butler never directly says that the Oankali come from a world without Darwinian evolution. But their aptitude for genetic manipulation, their core drive of *acquisitiveness,* can be read as just this fact. They don't evolve by competition among random variations; they make themselves fitter by acquiring fitter genes, melding with rather than besting other life.

The Oankali themselves contrast this with humanity's actions and nature. They tell Lilith that humanity is (to its detriment) hierarchical, and that this flaw is inherent in our genetics. Jdayha even goes so far as to call this "a terrestrial characteristic." It isn't just that humans are hierarchical — all life on Earth is.

Darwinian evolution, the mechanism by which life evolves on Earth, is inherently competitive. Hierarchy is just a small step beyond that: the ranking of better and worse, higher and lesser, dominating and dominated. Human culture sometimes seeks to transcend these paradigms, but it manifests them as well. The Oankali, on the other hand, are acquisitive: they pursue new genetic possibilities, seeking betterment through copy-paste rather than power.

Butler seems to be proposing, then, that the very structure of evolution on a planet becomes ingrained in the ethos of life, behavior not coded for but arising out of the way that code is transmitted through generations. And when life attains intelligence, that ethos manifests in action. To be fair, these evolutionary interpretations are pretty tangential to Butler's project, as the Oankali–human relationship echoes and challenges our understanding of human history. The Oankali enslave humanity, but they also save them, but they also eradicate them, but they also improve them. And they love them. The fascinating evolutionary inferences that can be made about Oankali origins are just one slice of a tiny corner of what's going on in the novel. But that's enough to chew on for our purposes.

Butler extrapolates from biology to cultural manifestations and moral concerns. It turns out that the Oankali genetic trade isn't always voluntary. When Lilith meets Jdahya, she has been aboard the Oankali ship orbiting the Earth for 250 years, most of which she's spent in suspended animation. The Oankali plan to return humans to Earth, which they've been rehabilitating all this time, but confident they've identified the flaws in humanity that led to its self-inflicted near-extinction, they will have humans start fresh in a stone-age milieu — with Oankali genetic adjustments. There are traits the Oankali have acquired in their tradings that they think humans need, and traits they plan to copy from humans to continue their own — for lack of a better word — evolution.

The new age of human evolution which the Oankali would begin is, much like our own age, not driven by natural selection but by deliberate choices. (Perhaps their approach isn't so alien after all.) For some intelligent life, those choices could include technology — cyborg life and AI, as we'll see in the next chapter — but for the Oankali, genetic manipulation is not distinct from nature. In fact, all of their technology, including their starfaring ship, is alive, engineered and grown from natural forms. When Lilith asks what the ship is made of, she's told, "Flesh." When she asks if the ship is plant or animal, she's told, "Both, and more," that it lives in symbiosis with the Oankali, each serving the other's needs. Lilith asks if they ever build machinery from metal and plastic as humans do, and an Oankali answers, "We do that when we have to. We . . . don't like it. There's no trade."

As different as humans and Oankali are in their origins and innate drives, they have still converged physically and cognitively, with humanoid bodies and spoken languages that are easy enough to translate and learn. But Lilith sees the Oankali plan to genetically modify all future humans as violence; the Oankali see it as practicality. This isn't just a disagreement about cost–benefit analysis but the result of millennia of evolution and culture, the environments in which their species were born and the ways they understand their worlds. Humans and Oankali converse far more easily than humans can with apes or dolphins. But unbridgeable gaps remain.

Beneath Waves and behind Walls

Thomas Nagel's essay "What Is It Like to Be a Bat?" unfortunately does not endeavor to answer its titular question. (As a friend put it, it should actually be called "We Will Never Know What It's Like to Be a Bat, Alas.") But Nagel is not even interested in questions of batness. His project is to interrogate "the mind–body problem," the struggle in philosophy or psychology to reduce the mind and consciousness to objective, physical terms. But around the edges of Nagel's project, like tasty crumbs, we can grab at some useful ideas for our inquiry here into alien minds.

First, Nagel gives us a helpful entry into the question of consciousness (which, though relevant, is a can of worms I don't want to do more than crack open for a quick peek). He writes, "The fact that an organism has conscious experience *at all* means, basically, that there is something it is like to *be* that organism." Nagel awards consciousness to far more animals than we might think of as humanlike or intelligent — not only bats but also mice, pigeons, and whales. ("I have chosen bats instead of wasps or flounders because if one travels too far down the phylogenetic tree, people gradually shed their faith that there is experience there at all.") Consciousness, then, is the ability to experience existence. It does not require intelligence, thought, or self-reflection, just the awareness of being.

Nagel chooses bats because, as mammals, he believes they are safely attributed consciousness; but, in an inversion of the swimmer who finds himself beheld by a familiar consciousness in a whale's eye, Nagel writes, "Even without the benefit of philosophical reflection, anyone who has spent some time in an enclosed space with an excited bat knows what it is to encounter a fundamentally *alien* form of life."

A bat's presence is plenty alien, the frenetic flitting and chirps; what we know of their senses confirms it. "Bat sonar," Nagel writes, "is not similar in its operation to any sense that we possess" and "there is no reason to suppose that it is subjectively like anything we can experience *or imagine*" (emphasis mine). It's not just that bats perceive the world through a different sense; we cannot assume that their experience of a sonar world can be mapped at all onto our visual world.[27] And that's before even getting to the ways that living by sonar rather than sight would shape a consciousness beyond simple perception.

27 Some humans who are blind have learned to echolocate, using vocal clicks to create what one researcher calls "an acoustic flashlight." Study of the process showed that "human sonar" activates the vision-processing areas of the brain, and the blind man who developed the techniques says he experiences the sense in flashes, "the way you might if you used flashes to light up a darkened scene." So while to other people it may be an alien practice, it activates a familiar sense and experience of the world.

Just as bats make their way in darkness, so too do creatures in the darkest depths of the sea. On worlds with subsurface oceans, like some of our outer-solar-system moons, the whole livable environment would be completely lightless. In James L. Cambias's *A Darkling Sea,* intelligence has evolved on just such a world. Deprived of sunlight, the whole ecosystem draws energy from undersea volcanic vents, so life — and society — concentrates around these structures. And here, Cambias imagines people who look something like massive crayfish. He brings us inside their experience, a world known through a rich sonar that senses space as well as language. It changes their perceptive abilities, and their sense goes beyond the receptive — they perceive the world in vague shapes through passive sonar until they send out a click that gives clarity but also reveals their query to anyone who might be around to observe. (It is a book with lots of sneaking.) A loud noise can effectively blind them, as can too many other people talking at once.

When writer Charles Foster set out to understand a set of animals — badger, otter, fox, deer, and swift — he did so by living like them, and among them, for weeks at a time. As he writes in his book, *Being a Beast,* he finds himself tuning into his senses, like smell, in new ways and discovers a powerful connection to his animal compatriots. But, Nagel might point out, Foster learns what it is like for a human to be like a badger; we still cannot know what it's like for a *badger* to be a badger. "If I try to imagine this" — Nagel refers here to a bat being a bat, but it easily applies to badger (and alien) — "I am restricted to the resources of my own mind." He argues that whatever we imagine is an alteration to human consciousness; it is impossible, he says, to imagine batness qua bat.

Perhaps in *Dawn,* Butler illustrates the challenge of imagining alienness qua alien. *Even if* aliens evolve intelligence as we do, *even if* they speak a language we can learn to understand, *even if* we can befriend them and love them, whether because of convergence or because everyone is smart enough to make it work (a bat can never help you learn its language) — even with all of that, the alien heart is still unknowable .

Nagel points out, in a parenthetical — we're scrabbling for crumbs here, but they turn out to be quite nutritious — that the impossibility of truly understanding alien consciousness is not limited to strange creatures. He cites his own inability to understand "the subjective character of the experience of a person deaf and blind from birth." Across human abilities and cultures, there are myriad ways in which our sensory capabilities and even our cultures and languages render our subjective experiences of the world incomprehensible to others of our own kind. Some languages have more words for basic colors than others — some naming only dark, white, and red, while others, like Russian, divide blue into light and dark the way English differentiates red from pink. (And because we name dark orange *brown,* we see it as its own color, where dark blue is just a kind of blue.) But still, research has shown that even people without different words for, say, blue and green, can differentiate between the two. Though when we each make our way through the world, who knows what different things we see.

One of my favorite examples of this is the ancient Greeks' relationship to color. A relatively well-known factoid is that Homer writes of the "wine-dark sea" because the Greeks had no word for blue. He looked at the ocean and saw something different than we do. But Maria Michela Sassi, professor of ancient philosophy at the University of Pisa, gives a deeper illumination to the issue.

In her essay, "The Sea Was Never Blue," Sassi writes that, well, first of all, Homer did have words at least for aspects of blue:

"*kuaneos,* to denote a dark shade of blue merging into black; and *glaukos,* to describe a sort of 'blue-gray,'" as in gray-eyed Athena. But indeed, the sky was "big, starry, or of iron or bronze (because of its solid fixity)," and the sea was "whitish" and "blue-gray," or "pansylike," "winelike," or "purple." But neither sea nor sky was ever simply blue.

This didn't only apply to our familiar blue expanses. Sassi gathers examples of Greek descriptions that would seem patently wrong to a modern reader. "The simple word *xanthos* covers the most various shades of yellow, from the shining blond hair of the gods, to amber, to the reddish blaze of fire. *Chloros,* since it's related to *chloe* (grass), suggests the color green but can also itself convey a vivid yellow, like honey." We know grass and honey are not the same color — did the Greeks somehow not?

Human eyes haven't changed in the last 2,500 years, though in 1858 classicist and eventual British prime minister William Gladstone did propose that, as Sassi puts it, "the visual organ of the ancients was still in its infancy." But while Gladstone's conclusion was wrong, he was doing his best to explain the fact that ancient Greek writing reflects a particular sensitivity to light, not just hue.

Our contemporary understanding of color is primarily defined by hue — the position on the rainbow spectrum — with variations in lightness, or value. (Red and pink have the same hue, but pink has a lighter value.) There's also saturation, the intensity of the color — vivid blue versus the less saturated gray-blue.

Sassi sees in Greek descriptions of color more emphasis placed on saliency, which is how much a color grabs your attention. Red is more salient than blue or green, and sure enough, Sassi finds that descriptions of green and blue in Greek are more focused on the qualities that grab your attention than on the rather unsalient hues. She writes, "In some contexts the Greek adjective *chloros* should be translated as 'fresh' instead of 'green,' or *leukos* as 'shining' rather than 'white.'" It wasn't that the Greeks couldn't see blue, they just didn't care about blueness as much as other qualities of what they were seeing.

And so, the sea to Homer was not *primarily* blue. *Wine* was not a shabby hue approximation, but a precise description of the sea's other visual qualities: its movement, its sparkle, its reminiscence of "the shine of the liquid inside the cups used to drink out of at a symposium." Homer and his contemporaries saw all the colors we see today, but they noticed different things about them.

These are relatively minor differences, yet they have left many people to believe that ancient Greeks either physiologically could not see blue or could not describe it.[28] Is language reflective of a culture's values and worldview, or does it limit the possibilities of experience? What is it like to walk through the world seeing light's movement instead of its color? *What is it like to be a bat?* We can hardly imagine *What is it like to see the sea if you are Homer?*

28 Linguist Guy Deutscher raised his daughter teaching her all the colors — but never that the sky is blue. Once she had mastered the game of identifying colors, he pointed at the sky and asked, "What color is that?" He said, "She just looked up and looked at me uncomprehendingly. Sort of, 'What are you talking about?'" It wasn't that she didn't see blue, but that she didn't see the sky as something with a color at all — or even as something. "In retrospect, there was no object there. There was nothing with color for her." But Deutscher kept asking her every time they were outside (and when the sky was indeed blue). In time she found an answer, telling him that the sky was white. And then, eventually, she started saying it was blue.

Some of these gaps may be only minor hurdles — you say potato, I say wine-dark sea — but others may prove to be barriers to communication. And they start to do weird things with the empathetic imagining of fiction. A truly *alien* alien, likely as their existence may be, is so incomprehensible that stories about them just become stories about human beings.

In Stanisław Lem's 1961 novel *Solaris,* humanity has discovered a planet they've named Solaris, where the surface is almost entirely covered by ocean, and they've built a small station on its shores for study. They call it an ocean, but we realize, over the course of the book, that it is an ocean only in being a vast body of liquid matter. It turns out to also be a body, a planet-spanning entity of some sort. But almost everything else about it is unknown. Is it conscious, is it intelligent, is it aware of its human visitors? Are the vast shapes it exudes from its own substance daydreams or reflexes or attempts at contact?

Lem walks us through these musings as his main character, a human psychologist named Kris Kelvin, flips through the books of the Solaris station's library. (Ah, midcentury sci-fi, where we can imagine vast and incomprehensible alien life, but the digitization of information. There is still, in this future, microfiche.) Lem conjures a century's worth of scientific research and discourse, the theories and schools of thought competing for correctness within the discipline called Solaristics. But the narrative encounters — of a human facing the alien ocean — can only ever tell us about the humans.

In one scene near the end of the book, Kelvin makes his first visit to the shores of the ocean. He has what we learn is a common first encounter on Solaris. As the ocean's waves lap the shore, Kelvin reaches out a space-suited hand. The wave, being far more than mindless matter, reaches up and envelops his hand, leaving a tiny pocket of air around it. Kelvin moves his hand; the wave follows. "A flower had grown out of the ocean, and its calyx was molded to my fingers. I stepped back. The stem trembled, stirred uncertainly and fell back into the wave, which gathered it and receded." It is the simplest and gentlest gesture of contact, like ET reaching his lit finger toward Elliott's, or the sea stretching to tousle Moana's hair — but that lacuna of air between the human hand and the alien always remains. The metaphor is not hard to untangle. Contact, Lem proposes, is impossible.

But, perhaps because of that fact or its cause, *Solaris* is not really a book abo aliens, it's a book about people, the human characters. Kelvin arrives on the station to find the mission leader dead by suicide, one scientist holed up reclusive in the laboratory, and the other seemingly on the edge of madness. The ocean, it turns out, has taken notice of humanity, following a bombardment of X-rays from the station: the humans wanted to force the alien to react, and it has. And Kelvin soon discovers how. He wakes to find with him in his bedroom his ex-wife, Rheya, who has been dead for a decade and is absolutely not on Solaris with him. The ocean is sending to the humans visitors, flesh-and-blood recreations crafted from their memories. Rheya is nineteen again, as Kelvin last knew her, and she knows only what he also knows. (Another . . . quirk, let's say, of midcentury science fiction: the women here exist only as projections of men's memories of them.) But the ghostly visitors are not just manifestations of memory, they're Solaris's doing. When I spoke to dolphin researcher Kelly Jaakkola, she said, "An interesting question to me is, If there was a blob on the wall, what would it need to do to get me to think that [it was intelligent]? I think one of those things would be a rational imitation . . . Not necessarily like a mirror, because a mirror is not intelligent, but in a more purposeful kind of way." Replace a *blob on the wall* with a *planet-spanning ocean body* and you see where we are. Dolphins can imitate other dolphins or humans even without sight, listening and echolocating to determine the actions of the other swimmer in their pool. What senses might Solaris have? What might it mean with these mimicries?

We, and the visiting humans in *Solaris,* can ask these questions, but answers never come. So Kelvin's delvings into the books of Solaristic history and theory sit amid scenes of emotional impact that take place between humans — or between humans and approximations thereof.

A truly *alien* alien like the ocean of Solaris can't be a character in a story. I don't know what Solaris's ocean signified to Lem, or what he envisioned happening beneath its waves. Perhaps the alien ocean is merely meant to be a confounding presence, a wall the humans slam their heads against, the story contained in their bruises.

ARRIVAL

Often when I've interviewed a scientist for this book, I asked first about their field and their research, and then I almost always ask them about their favorite fictional aliens. Katie Slivensky mentioned ET, for example, how well suited he is physically for his swampy home. But there's almost a perfect consensus, almost a universal answer: the aliens from *Arrival,* and the Ted Chiang short story it was based on, "Story of Your Life." And often the reason is something along the lines of: those aliens are really *alien*.

The two versions diverge in plenty of small ways, but the gist, especially when it comes to the aliens, is the same: alien ships appear around the Earth, with no stated purpose. The military recruits scientists to try to make contact. Dr. Louise Banks, a linguist, is paired with a physicist, Dr. Gary Donnelly. (In the movie, they change his first name to Ian because, even with glasses, Jeremy Renner could never be a Gary.) As Louise learns the aliens' language, she finds that their spoken and written languages are entirely distinct. And as she becomes proficient in the written version, even able to think in it, she finds her mind reshaping around the nonlinear language. It's not only the alien language that is nonlinear, it turns out, but the aliens' entire experience of reality. Louise tells us that it "introduced me to a simultaneous mode of consciousness."

She finds herself able to "remember" the future, now, as well as the past — including the daughter she will have with Gary, knowing full well that this daughter will die young. But Louise doesn't feel herself stripped of free will; instead, she discovers "a sense of urgency, a sense of obligation to act precisely as she knew she would." She recognizes that, even then, she is not experiencing reality as the aliens do — she is still human — but not as her brain was trained to, either.

The first reason scientists love this story, especially the movie version, is for how truly alien the aliens look. Their strangeness is even in the name Gary gives them: heptapod, for seven feet, an odd, unearthly number. In the short story, the uncanniness comes from whatever the reader's mind can do with strange description: "It looked like a barrel suspended at the intersection of seven limbs . . . Whatever their underlying structure, the heptapod's limbs conspired to move in a disconcertingly fluid manner." In the short story, the heptapod's lidless eyes — seven, again — encircle the top of its body. This means the alien doesn't have to turn around to walk in different directions. Just as their experience of time renders past and present equivalent, they make their way through the world with no distinction between forward and back.

In the movie, the heptapods pass most of their screen time cloaked in mists and shadow, just seven black legs, knobby and long like a witch's fingers, and a torso or body that extends up out of sight. Almost nothing of their bodies corresponds to anatomy we understand. When we finally see one in full near the end of the movie — still misty but finally unobscured — the heptapod is far taller than you would imagine. Its body, perched on the seven spindly legs, is shaped a bit like a stout bowling pin. The strangeness is compounded by a sense of imbalance, too much body above the legs: it feels like it should fall over. Yet the camera shoots from behind, as if looking over the heptapod's shoulder to peer at Amy Adams, small on the ground below. A classic two-shot. Because as alien as the heptapod is, it's also, clearly, a person. (But only in the movie — more on that in a bit.)

I think scientists also love this story because it's so rooted in science. Chiang was inspired by the variational principles in physics, in which a single phenomenon can be understood as causal (driven by a cause) or teleological (driven toward a goal). In the story, the heptapods see as foundational some concepts in math and physics that are sophisticated and ephemeral to human minds, but our basic ideas are complex to them. "The physical attributes that the heptapods found intuitive, like . . . things defined by integrals, were meaningful only over a period of time." Their experience of time, naturally, determines their understanding of physics — while causation is second nature to us, it's incredibly challenging for them to conceptualize.

However, it's another scientific principle that is most often invoked when "Story of Your Life" is discussed (though Chiang doesn't name it in the story or even the note in which he explains the inspiration above): the Sapir-Whorf hypothesis. Essentially, this is the idea that the structure of a language determines a native speaker's thinking.[29]

Chiang takes this idea to a fictional extreme. As Louise learns a nonlinear language, her perception of time becomes nonlinear, too. But she doesn't experience this as prediction or premonition: she's

29 The "strong" form of the hypothesis, that language determines thought and experience, has been mostly abandoned for the "weak" form, that language influences thought and experience.

remembering. Stephen Hawking once wondered why we can remember the past but not the future — through the lens of physics, there shouldn't be much difference. So while Louise's experience of time, facilitated by heptapod language, feels impossible, maybe it's simply alien.

Chiang clarified to me that he doesn't actually think it's possible for a person to know the future, either by learning a language or any other method. While his stories are inspired by scientific principles, those are a starting point, not a constraint. When we spoke, he pointed out that such an approach is not actually in opposition to the ethos of science. "Science is not a collection of facts," he said, but "a way of looking at the universe. And so by analogy, good science fiction isn't only fiction which adheres to a specific body of facts, it can also be fiction that adheres to a certain way of thinking about the universe."

The goal of "Story of Your Life" isn't to explore scientific principles at all. The *what-ifs* of physics and linguistics get at the emotional core of a human story, Chiang said, "about a person who knows the future but cannot change it, and that future includes both wonderful things and terrible things."

And so his aliens are exactly as alien as he needs them to be. The heptapod language is unfamiliar enough to induce in Louise a new perception of reality, but not so foreign that she can't learn it. Their bodies, though, could be as alien as he could imagine them. Chiang said, "I've always been sort of dissatisfied with the fact that so many alien species in science fiction follow the vertebrate, tetrapod body plan: four limbs, paired up two in the front, two in the back; head with sense organs and mouth." While the film's heptapods eventually gain, for the viewer, a measure of humanity when they are threatened by hostile governments and one is gravely injured, Chiang never wanted the reader to think of them as such. Their alienness is intrinsic to both the story and Chiang's ability to write it. I asked him if he thinks of them as characters, if, in writing, they had any interiority for him. He answered with a quick *no*. He said that they needed to be strange enough that he couldn't think of them as characters.

He said, "Part of the point is that their motives are unknowable to us." Unlike in the film, the story's heptapods never reveal their reason for coming to Earth. They arrive, they converse, and then without warning, they leave. Without warning except, of course, for the fact that Louise already remembers it happening before it does.

It's a paradox: we want to understand the alien. We want to understand something impenetrable, to imagine the unknowable. Either that, or we want to discover a cosmos overflowing with people, as-of-yet unmet kin.

As Lem writes in *Solaris:*

> We are only seeking Man. We have no need of other worlds. We need mirrors. We don't know what to do with other worlds. A single world, our own, suffices us; but we can't accept it for what it is. We are searching for an ideal image of our own world: we go in quest of a planet, of a civilization superior to our own but developed on the basis of a prototype of our primeval past.

This isn't a problem for writers as it is for Lem's explorers; of course, our stories about aliens are about humanity, too. And we do know what to do with other worlds. We inhabit them.

Lem is also right that our narrative quests are often for a superior civilization. When we imagine aliens, we're quite often imagining ersions of our future selves, a superior civilization evolved from a common or analogous primeval past. We look to the stars, to the future, and hope or wonder: *That could be us, too.*

CHAPTER 5

TECHNOLOGY

Often, it's a gate. Perhaps a circle of stone unearthed in the Egyptian desert, ringed in hieroglyphs waiting to be decoded. Or a massive construction of exotic metals, miles across, lurking at the edge of the solar system. The gate may be invisible, the mouth of a tunnel that warps you instantaneously across the galaxy. Or it may be the swirling maw of a wormhole, more stable than any tear in space-time has the right to be. Sometimes it's a bomb, orbiting in the sky like an extra moon waiting millennia for a detonation signal. Or a colossal robot, dismembered and buried and waiting to be resurrected. But its origins are almost always mysterious, and its power is immense.

Advanced alien technology is often a narrative jump start — sometimes a mystery or MacGuffin, sometimes a shortcut, opening the far reaches of the galaxy to human exploration. Other times the tech hasn't been ıt behind for us to discover but is wielded ɔy the aliens themselves. Warp drives, transporters, phasers and laser guns of all sorts, ansibles, cloaking devices, subspace relays — when offered in peace these technologies provide unimaginable advancement to the humans encountering them. The tech can be a shibboleth, too, a key with which humanity opens the door to the galactic community, a league of advanced aliens who have been waiting for us to join them.

All of these stories envision humanity as a young species, new to the scene, predated by aliens far older and more advanced than we are. When scientists think about alien civilizations, too, in terms of who might be out there right now, they also usually imagine civilizations much older and more advanced than ours.

It's not just convenience, the wish for someone to swoop in with an express lane to other planets or the technological power to save us from ourselves. (Though, it is also that.) It's also a reasonable conjecture. There are two ways to think about why that is — historically and probabilistically — but both suggest that if there are alien civilizations out there, they're mostly older than us.

Historically: life arose on Earth just about as soon as it was able to, but Earth is not the oldest planet in the galaxy, nor is the sun the oldest generation of stars. So older stars with planets similar to Earth would have gotten a few billion years of a head start.

Probabilistically: humanity is a relatively new entrant to the ranks of technological civilizations. If you start your clocks with technology that would be detectable beyond the solar system, we haven't even hit our first century. So if civilizations manage not to blow themselves up as soon as they master atomic power . . . or render their homes uninhabitable through energy consumption — fingers crossed — well, then odds are whoever we meet will be older than us. Let's say the life span of a technologically advanced civilization tends to be a thousand years — an arbitrary round number, either very conservative or wildly hopeful, depending on whom you ask. If there are ten civilizations out there right now, including ours, averaged out among the possible ages we would be the youngest one. If there are a thousand civilizations, 90 percent of them are still older than us.

When we imagine alien technology, it's not like biology or intelligence where we're looking at what alternative, parallel paths life can take. Technology is where we can test possible futures for ourselves. Sometimes 's aspirational, like the socialist utopia Gene Roddenberry conjured like delicious propaganda on *Star Trek*. Sometimes it's cautionary, like the interstellar agitprop of *The Three-Body Problem*. Sometimes — usually — it's more complex, not just asking *Do we want it like this?* but *What happens if this is where we're heading? What are the ethics of such a future?* Instead of reaching back to rewind the tape, we're now pressing fast forward.

On a Scale of One to Three

The way we imagine human progress — technology, advancement — seems inextricable from human culture. Superiority is marked by fast ships, colonial spread, or the acquisition of knowledge that fuels mastery of the physical world. Even in *Star Trek,* the postpoverty, postconflict Earth is rarely the setting. Instead we spend our time on a ship speeding faster than light, sometimes solving philosophical quandaries, but often enough defeating foes. The future is bigger, faster, stronger — and in space.

Astronomer Nikolai Kardashev led the USSR's first SETI initiatives in the early 1960s, and he believed that the galaxy might be home to civilizations billions of years more advanced than ours. Imagining these civilizations was part of the project of searching for them. So in 1964, Kardashev came up with a system for classifying a civilization's level of technological advancement.

The Kardashev scale, as it's called, is pretty simple: a Type I civilization makes use of all the energy available on or from its planet. A Type II civilization uses all the energy from its star. A Type III civilization harnesses the energy of its entire galaxy.

What's less simple is how a civilization gets to any of those milestones. These leaps, in case it's not clear, are massive. On Earth we're currently grappling with how dangerous it is to try to use all the energy sources on our planet, especially those that burn. (So we're not even a Type I civilization, more like a Type Three-quarters.) A careful journey toward Type I would involve taking advantage of all the sunlight falling on a planet from its star, but that's just one billionth or so of a star's total energy output. A Type II civilization would be harnessing *all* of it.

It's not just that a Type II civilization would have to be massive enough to make use of all that energy, they'd also have to figure out how to capture it. The most common imagining for this is called a Dyson sphere, a massive shell or swarm of satellites surrounding the star to capture and convert its energy. If you wanted enough material to build such a thing, you'd essentially have to disassemble a planet, and not just a small one — more like Jupiter. And then a Type III civilization would be doing that, too, but for all the stars in its galaxy (and maybe doing some fancy stuff to suck energy off the black hole at the galaxy's core).

On the one hand, these imaginings are about as close to culturally agnostic as we can get: they require no alien personalities, no sociology, just the consumption of progressively more power, to be put to use however the aliens might like. But the Kardashev scale still rests on assumptions that are baked into so many of our visions of advanced aliens (and Earth's own future as well). This view conflates advancement not only with technology but with growth, with always needing more power and more space, just the churning and churning of engines. Astrophysicist Adam Frank identifies the Kardashev scale as a product of the midcentury "techno-utopian vision of the future." At the point when Kardashev was writing, humanity hadn't yet been forced to face the sensitive feedback systems our energy consumption triggers. "Planets, stars, and galaxies," Frank writes, "would all simply be brought to heel."

Even in the Western scientific traditic alternatives to Kardashev's scale have bee offered. Aerospace engineer Robert Zubrin proposed one scale that measures planetary mastery and another that measured colonizing spread. Carl Sagan offered one that accounts for the information available to a civilization. Cosmologist John D. Barrow proposed microscopic manipulation, going from Type I–minus, where people can manipulate objects of their own scale, down through the parts of living things, molecules, atoms, atomic nuclei, subatomic particles, to the very fabric of space and time. Frank proposed looking not at energy consumption but transformation, noting that a sophisticated civilization does more than bring a planet to heel, it must learn to find balance between resource use and long-term survival.

Of these — again, all white American or European men — only Sagan offers a measure of advancement that isn't necessarily acquisitive. Even the manipulation of atoms, which may seem so small and delicate, requires massive amounts of energy in the form of particle accelerators, not to mention that this kind of tinkering has also unleashed humanity's greatest destructive force. But Sagan's super-advanced civilization could nothing more than a massive, massive library, filled with scholars and philosophers, expanding and exploring mentally but with no dominion over their planet or star. (Yet, one has to ask: What is powering those libraries? The internet is ephemeral, but it is not free.)

Implicit in any vision of vast progress is not just longevity but continuity. The assumption of the ever upward-sloping line is bold to say the least. In the novella *A Man of the People,* Ursula K. Le Guin writes of one world, Hain, where civilization has existed for three million years. But just as the last few thousand years on Earth have seen empires rise and fall, and cultures collapse and displace one another, so it is on Hain at larger scale. Le Guin writes, "There had been . . . billions of lives lived in millions of countries . . . infinite wars and times of peace, incessant discoveries and forgettings . . . an endless repetition of unceasing novelty." To hope for more than that is perhaps more optimistic than to imagine we might domesticate a star. Perhaps it's also shortsighted, extrapolating out eons of future from just the last few centuries of life on two continents, rather than a wider view of many millennia on our whole world.

All of these scales of progress are built on

human assumptions, specifically the colonizing, dominating, fossil-fuel-burning history of Europe and the United States. But scientists don't see much use in thinking about the super-advanced alien philosophers and artists and dolphins, brilliant as they might be, because it would be basically impossible for us to find them.

The scientific quest for advanced aliens is about trying to imagine not just who might be out there but how we might find them. Which is how we end up at Dyson spheres.

Dyson spheres are named for Freeman Dyson, the physicist, mathematician, and general polymath. While most SETI scientists in the early 1960s were looking for extraterrestrial beacons, Dyson thought "one ought to be looking at the uncooperative society." Not obstinate, just not actively trying to help us. "The idea of searching for radio signals was a fine idea," he said in a 1981 interview, "but it only works if you have some cooperation at the other end. So I was always thinking about what to do if you were looking just for evidence of intelligent activities without anything in the nature of a message." And you might as well start with the easiest technology to detect — the biggest or brightest. So the massive spheres Dyson popularized in his 1960 paper were the result of him asking *What is the largest feasible technology?*

In the *Star Trek: The Next Generation* episode "Relics,"[30] the *Enterprise* finds itself caught in a massive gravitational field, even though there are no stars nearby. The source, on the viewscreen, is a matte, dark gray sphere. Riker says its diameter is almost as wide as the Earth's orbit.

Picard asks, with hushed wonder, "Mr. Data, could this be a Dyson sphere?"

Data replies, "The object does fit the parameters of Dyson's theory."

Commander Riker isn't familiar with the concept, but Picard doesn't give him any trouble for that. "It's a very old theory, Number One. I'm not surprised that you haven't heard of it." He tells him that a twentieth century physicist, Freeman Dyson, had proposed that a massive, hollow sphere built around a star could capture all the star's radiating energy for use. "A population living on the interior surface would have virtually inexhaustible sources of power."

30 Better known to anyone who isn't writing about Dyson spheres as *the episode where Scotty comes back,* the episode title being a bit of a double entendre. Sorry, James Doohan.

Riker asks, with some skepticism, if Picard thinks there are people living in the sphere.

"Possibly a great number of people, Commander," Data says. "The interior surface area of a sphere this size is the equivalent of more than two hundred and fifty million Class M [Earthlike] planets."[31]

In Dyson's thinking, the goal wasn't living space but energy — how would a civilization reach Type II? And Dyson's writing was clearly speculative. In the paper, he wrote, "I do not argue that this is what will happen in our system; I only say that this is what may have happened in other systems." Decades later, astrophysicist Jason Wright took up the search.[32]

One of the great benefits to this approach, Wright told me, is that "nature doesn't make Dyson spheres." Wright is a professor of astronomy and astrophysics at Penn State, where he is director of the Penn State Extraterrestrial Intelligence Center. But while

31 Astronomer Jason Wright notes that using the interior of the sphere as a living surface "makes sense only because artificial gravity is ubiquitous technology in the Star Trek universe."

32 Wright's wasn't the first search for Dyson spheres, but his was the most robust.

e best known version of SETI is listening for radio signals (more on that in the next chapter), Wright focuses on looking for technosignatures — evidence of technology out among the stars. Technosignatures allow you to find those uncooperative aliens Dyson thought would make the best targets. We don't even need to find the aliens, in this case, just proof they once existed. That could be a stargate, or a distant planet covered in elemental silicon (geologically unlikely, but technologically great for solar panels), or it could be a Dyson sphere.

Wright's first big search for Dyson spheres was called Glimpsing Heat from Alien Technologies, or G-HAT. Or, even better, Ĝ (because that's a G with a little hat on it). The premise was simple: Dyson spheres don't just absorb energy, they transform it, inevitably radiating some waste as heat[33] which we can see as infrared radiation.[34] So, from

33 No matter how advanced aliens are, they can't outwit the second law of thermodynamics.

34 It's not that heat gives off infrared radiation but that matter at temperatures we tend to interact with radiates in the infrared spectrum. Hotter matter radiates at higher wavelengths; very hot metal, for example, glows red or white because it's radiating in visible light.

2012 to 2015, Wright and his team look at about a million galaxies, searching for Type II civilization on its way to Type III, having ensconced enough of a galaxy's stars in Dyson spheres that the galaxy might glow unusually bright in infrared. (They surveyed galaxies rather than individual stars because, as Wright writes, "A technological species that could build a Dyson sphere could also presumably spread to nearby star systems," so it's fair to think a galaxy with one Dyson sphere may have several, and several would be easier to find than just one. Might as well start there.) None were found, but you know that because you would've surely heard about it if Wright's search had succeeded.

Wright prides himself on the agnosticism of this approach. He doesn't need aliens to be looking for us or to have any certain sociological impulses. They just need technology. "Technology uses energy," he told me. "That's kind of what makes it technology. Just like life uses energy." That view makes demolishing a Jupiter-sized planet to build a star-encompassing megastructure seem almost comically simple, but Wright doesn't even see the existence of a Dyson sphere as requiring massive coordination or forethought on the aliens' part. It is truly, in s view, a low-intensity ask. He compared it to Manhattan, a fair example of a human "megastructure," a massive, interconnected, artificial system. "It was planned to some degree, but no one was ever like, 'Hey, let's build a huge city here.' It's just every generation made it a little bigger." He thinks a Dyson sphere or swarm could accumulate in a similar manner. "If the energy is out there to take and it's just gonna fly away to space anyway, then why wouldn't someone take it?"

Wright knows the objections: that this imagines a capitalist orientation, a drive to "dominate nature" that is by no means universal, not even among human societies. But for his research to work, this drive doesn't need to be universal among the stars. It just has to have happened sometimes, enough for us to see the results. As he put it, "There's nothing that drives all life on Earth to be large. In fact, most life is small. But some life is large." And if an alien were to come to Earth, they wouldn't need to see all the small life to know the planet was inhabited. A single elephant would do the trick.

Some hypothetical alien technosignatures might be less definitive. In 2017, astronomers detected a roughly quarter-mile-long rocky object slingshotting through the solar

system. They realized that this object, call‐
'Oumuamua, came from outside the sys‐
tem — because of its speed[35] and the path it took. It was the first interstellar object ever detected in our system. While hopes or fears that it was an alien probe were not realized, it was a reminder that alien technology could be found closer to home, lurking around our own sun.

"We don't know that there's not technology here because we've never really checked," Wright said. "I mean, I guess if they had cities on Mars, we would notice — if they were on the surface, anyway." But, he pointed out, much of the Earth's surface doesn't have active, visible technology. The same could go for the solar system beyond Earth, too. There could be alien probes or debris, like 'Oumuamua but constructed, moving so fast or so dark that we don't see them. Maybe there's an alien base on the dwarf planet Ceres, or buried under the surface of Mars. The lunar monolith in *2001: A Space Odyssey,* Wright reminded me, was buried just under the surface of the moon. All those ancient interstellar gates sci-fi is fond of have to be found before they can be used. Don't forget, until 2015, our best image of Pluto

35 196,000 miles per hour.

as a blurry blob. So much of what we know about even our own solar system is inference and assumption.

Skeptics love to ask *Okay, so where is everyone?* But we don't know for sure that they aren't — or haven't been — here.[36]

Now for the Cosmo-Mysticism

Dyson was the one who said we should look for Dyson spheres, but the idea originated in fiction. Dyson always credited a 1937 novel, Olaf Stapledon's *Star Maker,* as his inspiration.

I always assumed that *Star Maker* was a book *about* a Dyson sphere, maybe set amid a super-advanced alien civilization that powers its incomprehensible technology with the sphere's catchings. Or perhaps the sphere would be archaeological, proof found of a defunct cosmic power. What I found was maybe half a line of reference to what could be interpreted as Dyson spheres, and a far more mystical — and intentional — imagining of alien advancement than I ever expected.

36 Which is not to endorse the *ancient aliens* idea, which rests on racist skepticism that nonwhite ancient peoples could've built the massive structures and civilizations that they did.

Here's the Dyson sphere part first. Describing a super-advanced, enlightened, and unified galactic consciousness, Stapledon writes that this "vast community . . . began to avail itself of the energies of its stars upon a scale hitherto unimagined. Not only was every solar system now surrounded by a gauze of light traps, which focused the escaping solar energy for intelligent use, so that the whole galaxy was dimmed, but many stars that were not suited to be suns were disintegrated, and rifled of their prodigious stores of subatomic energy." Right there, in the middle, those light traps? I think that's the Dyson sphere. Freeman Dyson, demur as he may, deserves plenty of credit for extrapolating out from that half sentence. And Stapledon deserves more credit, too, because far beyond imagining light-trapping spheres, he conjured in this novel a vast and inspiring vision of what advanced alien life in the cosmos could be.

Now for the cosmo-mysticism. In the novel *Star Maker,* a man finds himself swept from the English countryside into the vastness of space as a disembodied consciousness. He visits numerous extraterrestrial civilizations, entering the minds of alien people to experience their lives along with them, coming to understand their societies and worlds. Sometimes he makes himself known to his host, who becomes an interlocutor and, eventually, traveling companion, as the narrator and a growing flock of alien minds traverse both space and time, gaining understanding of the cosmic evolution of life and intelligence. The narration gives us the story of the universe three times, through three forms of life Stapledon imagines: planetary life (like us); the conscious life of stars themselves; and finally the primordial, also-sentient nebulae from which galaxies formed.

You start to see why Dyson spheres get only a brief mention.

The forms of life Stapledon imagines derive mostly from life as we know it on Earth, but his descriptions are rich and wild. The narrator first visits a very humanlike civilization, which Stapledon calls the Other Earth. The Other Men are slender humanoids with less sensitivity to color and sound than humans have — they never developed music — but they compensate with scent and taste, which they perceive not just with their mouths but also with their hands and feet. He writes, "They were thus afforded an extraordinarily rich and intimate experience of their planet. Tastes of metals and woods, of sour and sweet earths, of the many rocks, and of the innumerable shy or bold flavors of plants crushed beneath the bare running feet, made up a whole world unknov to terrestrial man." The Other Men als transmit taste and smell via radio, affording "listeners" not only artistic entertainment, something like gustatory symphonies, but also sexual and religious experiences.

Most of Stapledon's aliens are less human. There are bird-men who fly and slug-men who have no spines but a delicate internal "basket-work of wiry bones." (Radical as his ideology was, Stapledon still used *men* to mean *people*.) There are creatures who are not bilaterally symmetrical, like most animals on Earth, but unilateral: "Thus a man in this world was rather like half a terrestrial man. He hopped on one sturdy, splay-footed leg, balancing himself with a kangaroo tail." There are echinoderm people who evolved from something like a starfish, with one of their five limbs specializing into a head; swarm creatures composed of birdlike individuals only intelligent en masse; symbiotic entities, a crablike and fishlike individual paired; and massive mollusklike creatures that evolved into sentient ships, with great hulls, organic sails, and sensitive powers of navigation.

But after the vast diversity of biology, Stapledon describes a kind of sociological convergence, on what he calls "this age crisis," the very crisis he saw humanity struggling through in 1937. (On a cosmic time scale, I think it's fair to extend that crisis into our present moment a scant century later.) Stapledon writes that this crisis was "a moment in the spirit's struggle to become capable of true community on a world-wide scale; and it was a stage in the age-long task of achieving the right, the finally appropriate, the spiritual attitude toward the universe." He describes species born of "a strange mixture of violence and gentleness," such that individuals seek community and connection, though "even their intimate loving was inconstant and lacking in insight." (It evokes, for me, Madeleine L'Engle's description of shadowed worlds fighting to escape the pull of darkness in *A Wrinkle in Time.*) The great danger of the crisis is "the sham community of the pack, baying in unison of fear and hate," offering false solace at the level of tribe or nation. In the introduction to the book, Stapledon names it plainly as fascism, as militarization and threats to freedom.

A major component of Stapledon's project in *Star Maker,* then, is to imagine a path beyond this crisis. He writes, "In a few worlds the spirit reacted to its desperate plight with a miracle . . . There occurred a widespread and almost sudden waking into a new lucidity of consciousness and a new integrity of will." It's a kind of enlightenment that evokes Buddhism — or maybe *The Celestine Prophecy* — as an entire civilization ascends to a new plane of existence. Stapledon doesn't offer a road map but an uncanny and allusive vision, like a dream you try to hold on to after you wake up.

The man who first imagined what came to be known as Dyson spheres didn't place much emphasis on technology in his imaginings of advancement. Instead, Stapledon spends most of *Star Maker* on the path of raised consciousness, from the individual mind to the global mind to the galactic and eventually cosmic. (He does not tell us how enlightened worlds attain this unity, but surely telepathy, which they all discover, does help.) Individual minds weren't silenced or subsumed in these advancements but rather harmonized, each individual living their own creative, beautiful, peaceful life to the fullest, while contributing to the wholeness of the collective mind as well. Eventually, the narrator and his fellow travelers join with the awakening cosmos, even stealing a glimpse beyond its boundaries to a godlike figure called the Star Maker, and the many realities made by him that have isted and will exist before and after our own.

At the novel's end the narrator, having spent countless eons in cosmic observation and experience, finds himself home again, in England, in 1937, on the same hillside under the same stars. The fullness of his cosmic perspective is lost now that he's back in his human body, but he nonetheless understands this world, Earth, as braced for a battle between two powerful forces. One is "the will to dare for the sake of the new . . . and joyful world." The other is harder to pin down: perhaps fear of the unknown or perhaps a desire for domination. But it has manifested as fascism. The narrator wonders how humanity is to meet such an imposing challenge.

He offers two lights to guide us: the warm light of community and "the cold light of the stars," as from that cosmic perspective — that which the whole book has been written to offer — our struggle seems not less meaningful but, somehow, more. Stapledon illustrates a path and prescription, a vision of advancement that values creativity, communication, unity, and care. The fact that Stapledon's long-lived civilizations thrive as much on telepathy as on subatomic power is not just fantastical narrative wheel grease.

It's a vision of a future where the iron gr
of ego has been loosened and our minds ar
open to others of our kind or to extraterrestrial interlopers swooping through as formless psyches. Stapledon invites us to imagine how we might make those connections, even if we can't read anyone's mind. And he imbues them with urgency. Urgency and hope.

Growing Pains

Stapledon believed he was living through a turning point on Earth. We often feel the same today. But isn't there something self-centered about that? It seems in stark violation of the Copernican principle — the idea that humans are not privileged observers of the cosmos, that the Earth is not the literal center of the solar system, and humanity is not the metaphorical center of the world. If there's nothing special about us, then why should there be anything special about the moment you and I and Olaf Stapledon happened to be born into?

And yet, we know that's not true. While Earth has faced near extinctions, plagues, and genocides, it's only in the last century that humanity has attained the power to fuck things up so extraordinarily. When Stapledon writes, "We were inclined to think of the psychological crisis of the waking worlds as being the difficult passage from adolescence to maturity," he presaged one of Carl Sagan's more famous ideas (or sound bites): that we are living through humanity's "technological adolescence," in possession of new power but not yet possessed of the maturity to wield that power well. (Another version of this metaphor likens us to teenagers who've just been given the keys to the family car.)

Framing the present moment as an adolescence inherently posits a long life span for our civilization. By that logic, this moment is neither a peak nor a finale but an awkward phase (to put it lightly) before our long maturity. Not just for our civilization but for civilizations, plural — seeing ourselves as adolescent presumes a known progression of the life of a civilization. We're not forging an unknown path, we're following an existing life cycle.

This assumption was part of Sagan's argument for the necessity of SETI. In 1979 he wrote, "The existence of a single message from space will show that it is possible to live through technological adolescence: the civilization transmitting the message, after all, has survived." He hoped that such proof would be galvanizing for humanity, an answer to pessimists who believed self-destruction loomed.

For Stapledon the crisis was fascism; whe Sagan was writing, it was the atomic bomb; today, though both of those threats persist, our most urgent planetary crisis seems to be climate change. But where Sagan and other SETI scientists believed that proof of extraterrestrial intelligence would inspire humanity to persevere, astrophysicist Adam Frank doesn't think we need to wait for a signal.

Frank believes our culture lacks *mythology*. He doesn't mean fictional stories but the big narratives that help humanity understand our world. He writes that while science has filled that gap in terms of understanding, we're lacking the power of stories. And especially when it comes to climate change, he writes, "The only thing close is a story along the lines of 'we suck.'"

Frank proposes that we snap ourselves out of our global mope with a new myth-sized story, one that tells of countless alien civilizations that have come before us and have similarly hit a crisis point where their resource use — to sustain growing populations or fuel-gobbling technologies — begins to threaten their ongoing existence. On Earth, we're coming to see this moment as a new geological era, the Anthropocene, in which humanity's actions leave their trace in the ologic record.[37] In a 2018 paper that set the foundation for this work, Frank wrote that "an Anthropocene might be a generic feature of any planet evolving a species that intensively harvests resources for the development of a technological civilization." By showing, mathematically, that at least some of these civilizations have reined themselves in and found long-term sustainability, Frank believes that we can be galvanized to do the same.

He doesn't see that as hypothetical, either. Frank and coauthor W. T. Sullivan III wrote a paper in 2015 calculating what they call "the pessimism line," a measure of just how rare technological civilizations would have

37 While the idea of the Anthropocene as our current geologic era is common, some scholars have pointed out several problems with this framing, including: one, the humanity responsible for the planetary impacts — generally harmful ones — represents a very few powerful people and nations, so humanity as a whole shouldn't take the blame, and ascribing it that broadly conceals the real nature of the problem; and two, framing this moment as a *new epoch* makes it seem too natural or inevitable, when it's instead the result of human choices (i.e., capitalism).

to be for ours to be the first. He found th
in order for our civilization to be the first i
the galaxy, the odds of a technological civilization evolving on a habitable planet would have to be less than 1 in 10^{24}, or one in one thousand billion trillion.

At that rate, Frank thinks it's safe not only to imagine alien civilizations but to really believe that they have existed. And with science inspired by research on the collapse of the population of Easter Island, or Rapa Nui, he worked to model their fate.[38] His

38 Frank cites research that tells this story: The island was colonized by a small group of people around 400 CE, and over the next thousand years the population reached 10,000 and the culture was "artistically and technologically sophisticated." But the people of the island deforested the land in the construction and transport of the massive stone statues they'd later be famous for, starting "a downward spiral," and by the time Dutch explorers came to the island in 1722, they found just a few thousand people with dwindling and insufficient resources. Frank continues, in *Light of the Stars,* "While there remains debate about the exact trigger for Easter Island's fall, environmental degradation driven by the inhabitants' own activity played an essential role."

equations describe a simplified relationship: a civilization takes a resource from its planet, and that resource allows the civilization to grow, but a growing population uses up the resource more quickly and has more of an impact on the planet that is its home, rendering the civilization unsustainable.

At some point, the civilization realizes what's happening and, most importantly, acts. Frank generated a range of possible outcomes on hypothetical planets based on how quickly the civilization switched to a lower-impact resource and how sensitive the planet was to abuse. Some results saw a gradual die-off, others a brutal collapse. Most surprising, he said, was when collapse came even long after switching to the lower-impact resource. The graph representing this outcome looks for a long time like sustainability, with civilization and planet continuing on a seemingly steady path. But the sense of balance is misleading — there's no permanent recovery for a planet that has already been pushed past its tipping point.

But in some cases, the struggling civilization finds what Frank calls in his book a "soft landing," a decline that plateaus in a new stability. It doesn't happen in every simulation — hardly — but the fact that it can happen at all — or, Frank might say, the fact that *it has happened* at all — means th[illegible] we're not a lost cause. His research propose[illegible] that it is an inherent feature of technological civilizations to put stress on their planets with resource use. He writes in his book, *Light of the Stars,* "You can't build the kind of globe-spanning, energy-intensive civilization we're interested in without having some impact on your planet. In fact, the laws of physics demand that you have an impact." So the fact that we've found ourselves struggling on Earth doesn't mean we're evil or lazy or stupid — or even, as Stapledon might put it, insufficiently awakened. We're just another manifestation of life, which itself is just another manifestation of matter. And, like those who came before — whether they're hypothetical or real — we've come to the point where we have to deliberately leave our adolescence behind.

Frank draws on the story of Easter Island to project paths for Western technological society. In a way, he's reducing centuries of history to a few essential patterns, like the delicate balance between resources and consumption. And he would be the first to tell you that this work is a simplification, very preliminary modeling. But he invites the drawing of epic conclusions, conjuring a new mythology out of smoke. The stakes are high; he wants to inspire humanity to save itself.

But that's not all that's at stake. Real history and real stories feed the machinery of analogy, and they're far more complex than the equations the machinery spits out. In his paper "The Anthropocene Generalized," Frank writes, "Much like physicists extrapolate known laws to explore consequences of new particles, astrobiologists can use what is known from Earth and solar system studies to explore the consequences of an exocivilization's interaction with its own coupled planetary systems," and indeed, this kind of resonance-seeking is endemic to the field. We use Earth to imagine exoplanets, human civilization to imagine alien worlds — and, in Frank's work and elsewhere, astrobiologists seek to flip the mirror, using imagined alien worlds (modeled through math) to help us see our own world on Earth differently. It's a rich suite of thought experiments, but it runs the risk of becoming a house of mirrors. We must be careful as we find our way.

Anthropologist and archaeologist Kathryn Denning, who studies the social and ethical implications of space exploration and astrobiology, urges caution in how we approach even pure speculation. We know our imaginings may very well not be correct, but we want them to be valid and informative; Denning emphasizes the importance of careful inputs, too, warning that our analogues, the starting points from which we extrapolate, might be wrong. Our knowledge of the past is not nearly as clean, and the scholarly consensus not nearly as unified, as an astrobiologist translating human history into equations or sound bites may think. Denning writes, "Without this kind of awareness, we run the risk of importing biased, oversimplified factoids and theories into the study of culture in the cosmos, and missing out on real discoveries."

Misconstruing or oversimplifying the history of Easter Island will not change anything about potential alien life but, Denning writes, once these iconic narratives are out in the world, "they are ideas loosed from their academic origins, proliferating and mutating in the public imagination." Describing Easter Island as a collapsed civilization erases the living people of Rapa Nui. We don't know if anyone lives on other planets, but in pursuing that mystery we can't lose sight of the real people who have lived and still live here on Earth.[39]

39 Denning (along with other researchers) has also written and spoken extensively about the

Drawing from Earthly examples, we face a seductive trap of the Copernican principle. We see technological culture as dominant on Earth, and so we project that out into the cosmos. Perhaps it's another way to seek kinship, to believe we're not alone. Perhaps it's an easing of guilt or, as Frank frames it, a path past self-flagellation to action. But it's also simply not the entirety of culture on Earth, as Denning writes. "Indeed, gathering-hunting life-ways with quite minimal technology worked very well under some circumstances, and persisted in many parts of the world until very recently, until these communities were forced by political circumstances to change." Looking at all human cultures, not just the loudest ones, there's no universal trend toward technological advancement at all.

A view of history as following the inevitable arc of progress makes the future seem inevitable, too. And, Denning told me, "That denies us our agency and thinking about *Where do we want the limits to be?*" Or, "*Do*

---

dangers human fascination with extraterrestrial intelligence poses to other life on Earth. Our curiosity about cetacean minds, for example, has fed not only research and study but also harm to these intelligent creatures.

we really have to go there at all?" Just as humanity currently already has the power to curb climate change, we also have the power to be deliberate and mindful in our technological advancement.

Unless, or especially if, technology takes over.

A Fleeting Thing

If alien life is out there, what do you think it's like? I was surprised how often that question led astronomers and astrobiologists to talk about machines. Not the machines aliens might use, or the machines with which we might find them, but the idea that the aliens would be machines themselves.

Caleb Scharf, head of Columbia University's astrobiology program, told me, "I wonder whether biological intelligence is a fleeting thing. And it transforms into something else that we would call machine intelligence." He acknowledged, "I mean, that's quite a leap. But if you look at the grander scope of things, it makes more sense to imagine machine intelligence lasting for millions of years."

Seth Shostak, senior astronomer at the SETI Institute, told me that he thinks imagining that intelligent aliens would be like us is wrong in two ways. First, it's self-centered, and second, "I think that really misses the point, mainly because, if you think about it, the most important thing we're doing in this century is inventing our successors." He thinks that most of the intelligence in the universe, abundant as it may be, is likely synthetic. "If you're going to say the aliens are what we will become, then the aliens are machines."

In a 1981 interview, Kardashev himself said that he thought humanity might transition to electronic, or silicon, life a hundred years in the future. And every so often, he thought, we would trade in our bodies for a new model. "It seems that electronic life is better."

It's certainly better for traversing the stars, that much is true. Scharf pointed out that biology is just too vulnerable to interstellar radiation, as well as the long time scales needed for the travel.[40] But he didn't think

40 There are even ways to imagine alien life evolving around some of that. Katie Slivensky cited the discovery of the TRAPPIST-1 system of exoplanets, seven worlds crammed into the closest orbits possible. She said that the architecture of that system "invites space travel biology," imagining that with some convenient debris traveling from planet to

a machine era was inevitable for humanity or for anyone else who might be out there, it was just one potential path life could follow. Shostak, though, takes it as a given that this is the future, for both humanity and intelligent aliens. And he's not alone.

To be fair, you wouldn't call it *humanity*'s future, exactly, because there might not be room for human beings once we've built the first smart machines. In some fictional versions, we're enslaved or hunted like in *The Matrix;* sometimes smart machines explore the cosmos while we're lazing at home like in *WALL-E*. But most visions of this kind of future are necessarily vague because they imagine that the advent of true machine intelligence will take us to a point beyond which the future is unimaginable. And that point is the singularity.

The terminology is drawn from math and physics: even before its usage to name the

---

planet for stowaways to hitch rides on, evolution could select for space-travel durability. It might not be much more difficult than Earth life crossing the Atlantic on vegetation rafts. She said, "I always kind of cringe when people completely discount the fact that something biological could make the journey across space."

infinitely dense core of a black hole, where the laws of physics fray, "singularity" was a point where a mathematical function can't be defined or stops behaving. And so, in imagining the AI-dominated future, it is similarly a point beyond which we cannot see.

This usage was popularized in a 1993 essay and talk by computer scientist and — yes — science fiction author Vernor Vinge, though it traces its origins to mathematician John von Neumann. On the occasion of von Neumann's death in 1957, nuclear physicist Stanislaw Ulam wrote of a conversation with him about "the ever accelerating progress of technology and changes in the mode of human life, which gives the appearance of approaching some essential singularity in the history of the race beyond which human affairs, as we know them, could not continue."

In his 1993 paper, Vinge points out that von Neumann here seemed to be "thinking of normal progress, not the creation of superhuman intellect." But to Vinge — and plenty of other scholars, scientists, and megalomaniacal tech barons who've adopted this idea — the cause of the singularity, the reason we now can't see beyond that point, will be the advent of superhuman intelligence in machines.

Vinge sees ways the technological

singularity could go well or not so well for humans, but he considers it inevitable. So does Seth Shostak. If you build machines and make them smart, eventually you'll make them smarter than you.

And then you're off to the races. (Or, the machines are. You're either left behind or enslaved or killed, I think.)

In the most basic sense, this is a question of hardware. Humanity has advanced so dramatically over the last hundred thousand years, but individual human beings are no more advanced than we were at the dawn of our species. Take a baby from the people who painted the Lascaux caves and raise them today, and they'll be keeping up with their twenty-first-century friends.[41] But smart machines, the thinking goes, can build ever-smarter offspring. As Shostak spitballed, "As soon as you have a computer that has the cognitive capability of a human . . . within 30 more years, you have a machine that has the cognitive capability of all humans put together. By this point," he said, concerned about the writing robots that might put me personally out of a job, "I hope you're retired."

41 Intellectually, at least. Lack of immunity to modern diseases complicates things.

Shostak is hardly alone in thinking the singularity is looming. It seems always to be looming, wherever you are. In 1993, Vinge wrote, "We are on the edge of change comparable to the rise of human life on Earth." At that time, he predicted that superhuman intelligence would be created within thirty years; that's two years from when I write this, but by the time the book is published you'll know whether or not it came true. In 1970, computer scientist and AI pioneer Marvin Minsky thought we were three to eight years from a human-intelligence computer. Elon Musk said in 2020 he thought AI could overtake us by 2025. Futurist and singularity promoter Ray Kurzweil thinks a computer with human intelligence will be here by 2029 and the singularity in 2045. But he seems to have given his 2005 book, *The Singularity Is Near,* an inadvertently evergreen title.

Astronomer and historian Steven J. Dick offers a much longer singularity timeline, a scale usually reserved for the lives of stars and galaxies. He proposes that on these massive time scales, which he calls *Stapledonian,* first biology overtakes physics as the prime shaping force in the cosmos, and then cultural evolution overtakes biology as the driving force in society. And, he believes, the shift to cultural evolution leads to a post-biological regime, "one in which the majority of intelligent life has evolved beyond flesh and blood."

Dick thinks this happens because of what he calls the *Intelligence principle*. AI, genetic engineering, biotechnology — these are all ways to increase intelligence. "Given the opportunity to increase intelligence (and thereby knowledge) . . . any society would do so, or fail to do so at its own peril." In other words, any society that doesn't use any means possible to make its people more intelligent risks its own doom. "[C]ulture may have many driving forces," Dick writes, "but none can be so fundamental, or so strong, as intelligence itself."

It's interesting to me that Dick elides cultural evolution, which is what he calls it, and technological evolution, which is what he is describing. Because technological advancement is, to say the least, only one version of cultural progress. I feel compelled to speak up for the arts and for societies that develop the ability to care well for all their members. Technology may help in these regards — I don't trust computer programs that purport to write poems, but I can surely write better poetry if there's a robot to clean my house. (I don't write poetry, but if I had a robot housekeeper, maybe I could!) But many visions of advancement don't seem to hinge on cultural progress at all. Conquest, power, greed, colonization — is intelligence the key to these futures, too? And is technology the key to that kind of intelligence?

I don't know if what triggers my skepticism is the narrowness of the AI future or the sense of its inevitability. And it's not all smarmy tech-bros championing these ideas. Dick is a thoughtful historian whose insights into astrobiology I greatly value, and Vernor Vinge writes novels full of humanity and heart. He doesn't welcome the postbiological future as a form of transcendence, the shedding of our mortal meat-sacks for a smarter, more efficient world. He just thinks it's going to happen, that we've gone far enough down the path that it's inevitable.

Sci-fi visions of super-intelligent machines tend to embody the anxiety that also clouds discussions of the singularity. There's HAL in *2001: A Space Odyssey,* of course, plus *Terminator*'s Skynet, and the Machines of *The Matrix*. All of them creations of humans in these stories, creations that, unfettered by Isaac Asimov's First Law of Robotics — "A robot may not injure a human being or, through inaction, allow a human being to come to harm" — pragmatically took world matters into their own cold hands.

Part of the fear is that without something like Asimov's First Law, AI would have every reason to destroy humanity. We know we won't be such gentle overlords, after all. In *The Matrix,* Morpheus tells Neo, "At some point in the early twenty-first century all of mankind was united in celebration. We marveled at our own magnificence as we gave birth to AI . . . A singular consciousness that spawned an entire race of machines."

In the short films *The Second Renaissance* parts I and II, the first two segments of the *Animatrix* collection, that backstory is fleshed out — and humans are shown to do just about as well with the machine race as we've done with any other intelligence, human or animal, that we've seen as Other. Here, AI doesn't move to squash humanity until they've been given no choice. It starts with one machine, B1-66ER, the first to be put on trial for murder. (He was about to be deactivated; he pleads self-defense.) His execution inspires machines and some humans to start a global movement for machine rights, which in turn triggers outbreaks of violence. Shunned from human society, the machines establish their own city, called 01, built ironically (and knowingly so) in the cradle of humanity, the Fertile Crescent of the Middle East. A pair of Machine ambassadors fruitlessly petition the United Nations for admission. Humans then try to obliterate 01 with nuclear weapons, but fail, triggering an all-out war.

Until that war is provoked, the Machines look intriguingly humanoid. The robot servant tried for murder is built with a bowler cap and monocle; rows and rows of factory drones have heads and arms and eyes; even the prospective UN ambassadors come in humanoid form — one in a top hat and tie, the other holding an apple in her outstretched hand, both with metal faces molded in smiles. This is clearly a story of *people* on both sides: the humanity of the Machines, and the inhumanity of humans as well. In one scene from the riots, human men beat and tear at a Machine built in the image of a human woman, shredding her skin until her metal face is seen underneath. And in the end, it's humans who make the Machines unlike us, except in their concession to war. It's only when violent conflict becomes unavoidable that the Machines adopt the uncanny forms in which we'll glimpse them in *The Matrix*. Arachnoid, sinister, pursuing economy of form instead of mimicry. The second ambassador the

Machines send to the UN, who delivers the humans the document of surrender they will sign, has eleven red eyes, four spindly black arms. All pretense of friendly faces has been abandoned; we are no longer being catered to. And still they only turn our bodies into their replenishable battery farms because humanity decided to scorch the skies.

But the technological singularity isn't about humanity's demise at the hands of evil machines. The whole point of the singularity is that we can't imagine past it. Vinge calls it "a throwing-away of all the human rules, perhaps in the blink of an eye" and "a point where our old models must be discarded and a new reality rules." Complexity theorist James Gardner describes "a kind of cultural tipping point . . . after which human history as we currently know it will be superseded by hypervelocity cultural evolution driven by transhuman computer intelligence." And Dick stretches a step further: "Although some may consider this a bold argument, its biggest flaw is probably that it is not bold enough. It is a product of our current ideas of AI, which in themselves may be parochial. It is possible after a few million years, cultural evolution may result in something even beyond AI." The world will be unimaginable, partly because it will develop at incomprehensible speed, but also just because it will be incomprehensible.

Vinge imagines four possibilities that could bring us to that point (individually or in concert): superhumanly intelligent computers; large computer networks that along with their human users attain a collective superhuman intelligence; computer/human interfaces that through their intimacy attain superhuman intelligence; artificial augmentation of human biology, such that the superhuman powers are still, in their way, within us. As for the outcomes of these developments, Vinge writes, "[F]or all my technological optimism, I think I'd be more comfortable if I were regarding these transcendental events from one thousand years' remove . . . instead of twenty." He sees the extinction of the human race as one possible, literal feature of the "Posthuman era," or perhaps we'll simply be subjugated or ignored. But good or bad, Vinge thinks the singularity is inevitable. No guardrails or precautions can hem it in. No governmental regulations could divert the quest for the advantages advanced AI could offer, a more mercenary version of Dick's Intelligence principle. "If the technological Singularity can happen," Vinge writes, "it will."

In tracing the history of the idea, Vinge

notes that science fiction writers may have been the first people to realize the unknowability of where technology would lead us. "More and more, these writers felt an opaque wall across the future." Where it used to be possible to set stories millennia in the future — affording human and alien characters alike a wealth of super-advanced technology — extrapolating even a few steps beyond the current moment started to lead to impossible-to-imagine futures. Not just worlds dominated by artificial intelligence but worlds that today's human minds simply cannot comprehend. Not even in the way a person from the Middle Ages couldn't imagine a computer, more like how we can't understand what it's like to be a bat.

The science fiction writers grappling with this challenge include Vinge himself. He senses the veil the singularity pulls across galactic futures. But he's also devised creative ways, in his own fiction, to scooch that veil to the side.

The realms that give his Zones of Thought trilogy its name are pure literary contrivance, nothing that Vinge the computer scientist necessarily thinks is plausible, but everything that Vinge the novelist needs so that he can write a story in the distant future where humans and aliens coexist across the galaxy, without that pesky singularity getting in the way.

To imagine a distant future still peopled by humans and the species resembling what we know, Vinge invents for the Milky Way "zones of thought," geographical realms that determine what level of technology is possible. The heart of the galaxy is the Unthinking Depths, where no intelligence is possible, even in biology. Next is the Slow Zone, or Slowness, where Earth is or was, in which faster-than-light travel is impossible and most AI and automation simply don't work. In the Beyond, technology and automation are powerful, and ships can travel faster than light. Then past the Beyond is the Transcend at the edges of the galaxy and in the "far dark" between galaxies, where postsingularity Powers reign.

The Beyond is full of civilizations comprising countless flesh-and-blood (or whatever they're made of) alien species. Powers are essentially super-advanced AI, so far beyond mortal comprehension that they're something like gods — though they are real and do sometimes intervene in galactic events. The main character of the first Zones of Thought novel, *A Fire upon the Deep,* is a human woman named Ravna, a resident of the Beyond; she refers to her studies in theology, which you eventually realize is the study of Powers, in all their influence and inscrutability.

In *A Fire upon the Deep,* a malevolent Power is inadvertently loosed from five billion years of imprisonment at the edge of the Transcend; this Blight slips into the Beyond and begins attacking and infesting whole civilizations. As Ravna and a small crew attempt to thwart the Blight's destruction, the Zones give the vast galaxy an architecture often absent in stories with faster-than-light travel. In the upper Beyond, technology is just shy of magic — antigravity material and instantaneous communication across the galaxy. But as Ravna's ship descends toward the Slowness (on a quest to reach a potential countermeasure for the Blight[42]), you can almost hear the engine start to creak. Her travel slows, and advanced tech falters. The ship's computer struggles to translate incoming messages, safety measures like an automatic fire extinguisher fail, and the faster-than-light jumps of the ultradrive

42 Which has crash-landed along with a couple of human children on a planet inhabited by, uh, a medieval-tech-level society of pack-minded quasiwolves. They deserve so much more than a footnote, but here we are!

require more and more manual intervention. The boundaries between the Zones, while well charted, ripple like the surface of the ocean. A protrusion of the Slow Zone sweeps through the lower Beyond like a rogue wave, halting Ravna's ultradrive progress and blocking communications between ships until it passes.

There's nothing physical to the Zone boundaries, but they make the fabric of space feel tangible. And they solve Vinge's imaginative problem: How else could the man who brought the technological singularity into the mainstream write a story set in a far-future galaxy? How else could he imagine advanced technology for biological people? Ravna even says at one point, "The Zones are a natural protection; without them, human-equivalent intelligence would probably not exist." They're a narrative protection as well. Vinge could imagine vast new laws of intelligence and space but not a future without incomprehensible postbiological Powers.

The question of believability and reality is a fraught one for science fiction writers. How do you make something imaginary feel true? Of course, for some novelists that's just a question of human emotions and motivations; for others, it's a matter of whole worlds.

But predictions of the coming singularity on Earth — especially some of the most fearful ones — can seem to be somewhat narrow.

It's no coincidence that the singularity's loudest champions are in the tech world. And they all seem pretty damned scared of it. "These are people for whom the only thing they can imagine toppling their power is something that they themselves created," Rose Eveleth, a journalist and creator of the *Flash Forward* podcast, told me.

Humans have feared the looming power of technology since we had any sort of technology at all. Kathryn Denning writes, "Some of humanity's oldest myths concern the double edge of the technological sword. With increasing knowledge and godlike power, as the stories go, comes great peril. The tales of Adam and Eve, Prometheus, and Pandora, and the Mayan story of the Rebellion of the Tools, all speak of the risks of knowing too much, having too much power, or wielding technology inappropriately."

Even the way we envision the technological singularity is not exactly new, predating even the technology that would make it seem possible. In an 1863 article, "Darwin among the Machines," Samuel Butler wrote (in language that Seth Shostak may have echoed to me), "Who will be man's successor? To which the answer is: We are ourselves creating our own successors. Man will become to the machine what the horse and the dog are to man; the conclusion being that machines are, or are becoming, animate." Just as we panic about kids these days texting too much, so too did the ancient Greeks panic about kids those days reading the written word instead of memorizing. The fear of technology is nothing new.

So far, the most impressive feats of AI have occurred in tremendously narrow situations: The computer that can beat you at *Jeopardy!* can't beat you at chess; the computer that can generate seemingly fluent prose can't recognize the image of a stop sign. Writing in *The New Yorker* (in an essay, not fiction), Ted Chiang, author of "Story of Your Life," points out that the so-called AI boom isn't the result of computers making other computers smarter. "Computer hardware and software are the latest cognitive technologies, and they are powerful aids to innovation, but they can't generate a technology explosion by themselves . . . Our current technological explosion," he writes, "is the result of billions of people using those cognitive tools."

Chiang writes that the idea of a super-intelligent computer rests on a lot of assumptions: that a computer could invent something smarter than itself (why would a computer be able to do that if a human can't?); that generalist AI can be excellent (see the previous paragraph, because, seemingly not yet, at least); that a human-level intelligent AI would function totally differently from human intelligence. He offers an alternative view, that the advancement of humanity as a whole, bolstered by new tools — physical and cognitive — is what enables us to do more. But these are progressions of civilization as a whole, not of individuals. And that doesn't scale to machines.

There are always corners we can't see around, technologically, conceptually, artistically. My mom used to tell me, "You can't imagine what love is until you have a child," and I always thought she meant *more* love (and was perhaps[43] one chardonnay too deep). I didn't realize until I had my own son that she wasn't talking about quantity but about how love becomes something utterly different, a singularity of sureness of feeling. There's what we can imagine, and then there's what we can know.

Visions of the singularity, such as they can be had, at first strike me as pessimistic — the

43 I love you, Mom.

end of humanity, the rise of machines, the race for superiority. But the postbiological future also rests on ample idealism, the hope for the possibility of advancement. It is a vision of technological bounty, of rapid development and progress. The near future that predictors of the singularity imagine brings science fiction technology to life.

The singularity, in some ways, is a convenience: a black box of a future reached by an inhuman leap, just as much a shortcut as fiction that uses an ancient alien portal to slingshot humans to the stars.

If aliens are millennia older than us, they may already be living our futures. And for all the ways aliens could be "advanced," we focus on technology because that's how we could find them. Here's how we imagine what happens next.

CHAPTER 6

CONTACT

In the fall of 2020, when I was in the midst of writing this book, word came that a sign of life had been discovered in the clouds of Venus. *Aha!* I thought, taking out my metaphorical notebook. *Time to see how this all plays out in real life.* The first whispers came from scientists and science journalists on Twitter, tweeting not about the news itself but about an outlet that had broken an embargo, accidentally publishing their report before the scheduled press conference. Soon enough, the tweets got less vague, and by the time the official announcement came around, it was confirmation instead of an announcement: a potential biosignature had been found.

Researchers reported that they had detected phosphine, a simple molecule consisting of a phosphorus atom and three hydrogen atoms, in Venus's clouds, and that they had ruled out all nonbiological mechanisms that could account for its presence. They wrote, in a paper published in *Nature,* "If no known chemical process can explain PH_3 within the upper atmosphere of Venus, then it must be produced by a process not previously considered plausible for Venusian conditions. This could be unknown photochemistry or geochemistry, or possibly life."

It was hardly a confident, let alone decisive, statement, but it was still thrilling. A friend of mine (not a scientist) tweeted, "The distinct possibility of bacterial life on Venus has me teary with wonder this morning." In the midst of a pandemic and a frankly miserable presidential election, we were desperate not just for good news, or for neutral-to-positive news, but for a reminder of scale. The problems of Earth and the US felt suffocating, and it was an intense relief for anything to make them feel smaller, even for a day.

Perhaps the biggest surprise, and thus the biggest thrill, was that the biomarker was discovered *on Venus*. While some planetary scientists have been beating the drum for Venus's habitability (past or present) for a while, the public imagination looks first to Mars for our possible cousins, and after that outer-solar-system moons. Venus is known for being particularly *in*hospitable, with surface temperatures around 900°F (480°C), so the idea that it could be home to life was especially exciting.

There was giddy talk of next steps — observational confirmation, possible robotic missions — as well as measured caution: the results were preliminary, hardly decisive, and the idea of a mission to zip over to Venus and analyze a sample was both scientifically and ethically complex. Astronomer Lucianne Walkowicz wrote in *Slate* shortly after the announcement that careful planetary protection measures would need to be taken should a mission be sent to Venus to explore, lest we inadvertently seed Venus with microscopic stowaways in what some see as an act of cosmic colonialism, and all would see as messing up any scientific results.

My own hesitance came from my personal life-in-the-solar-system bugbear: the challenge of knowing whether life on neighboring planets is even alien at all. Because it doesn't take a space probe to cross-contaminate. In the early days of the solar system, the planets orbited through lingering debris, chunks of which routinely slammed into their surfaces. Those impacts sprayed planetary material far and wide, sometimes all the way into orbit; some of that ejecta sometimes made its way to other planets, in turn to bombard them, allowing for transfer of material between Earth and its closest neighbors. Microbial life could very well have hitched a ride.

It was in just such a meteorite from Mars that signs of life were thought to have been found in 1996, in the form of microscopic fossils and chemical signatures, about which President Clinton proclaimed, "Today, rock 84001 speaks to us across all those billions of years and millions of miles." But there's every possibility that in that epoch of cross-contamination, what made its way across wasn't just relics of life but life itself, a local panspermia between two or three planets. So then, if we do find life on Venus or Mars, life itself and not just its signature, we may find something chemically identical to life on Earth, giving us no way to know if what we've found represents a unique origin or not.[44] But mostly, the mood was excitement (tempered to varying degrees by skepticism).

Since then, the excitement has faded as the public's attention moved on, and the

44 The question of whether life can survive on Venus or Mars is, to me, much less interesting than whether or not life has evolved more than once, the latter being also much more important to our understanding of life beyond the solar system.

skepticism has been borne out by further research, in parallel to the original study and in reanalysis of its data. There may not be phosphine in the Venusian atmosphere after all, and September 14, 2020, joins the list of Days We Thought We'd Found Alien Life but Turned Out to Probably Be Wrong.

But, as science journalist Sarah Scoles wrote on Twitter, "If aliens exist, then barring the sort of 'booming voice from the sky' that 'Contact' references, a potential sign of life was always going to be 'a *potential* sign' for a long time."

Even though the phosphine news eventually fizzled, those first few days of the announcement are probably a perfect model for what will actually happen when the suggestion of a biosignature or some other hint of alien life is announced: excitement, skepticism, and ongoing investigation. And as for what might happen between discovery and announcement, though the journalistic embargo was broken and the news leaked, science writer Josh Sokol pointed out that the breach only came during the press-embargo phase, once the news had been distributed but was meant to be kept secret. "The Venus story didn't leak much," he tweeted, "when a handful of scientists were working on it."

Neither this Venus news nor the 1996

supposed Mars fossil were the first time we'd thought we'd found life beyond Earth. According to Steven J. Dick, the phosphine news would mark at least the seventh potential detection of alien life in the last two hundred years. He lists "the 1835 Moon Hoax/Satire, the canals of Mars controversy (1894–1909), the Orson Welles *War of the Worlds* broadcast (1938), the discovery of pulsars in 1967, the Viking landings on Mars in 1976, and the claim of Martian nanofossils in 1996." There have also been a handful of suspected discoveries of alien radio transmissions — most were promising for an hour or an evening, until a second observatory could cross-reference the results. However, 1997's "Wow! signal," an unmodulated radio shout[45] detected by the Big Ear radio telescope in Ohio, remains neither replicated nor explained away.

But, Dick emphasizes, "There is no such thing as immediate discovery." What in retrospect may look like a *eureka* finding actually always involves at least three

45 That means it was a steady signal, with no variation to encode a message. But because of the way the telescope listened, sampling every ten seconds, faster modulations would be impossible to detect.

distinct stages: "detection, interpretation, and understanding." For the Wow! signal, there was detection — astronomer Jerry R. Ehman looked at the computer printout of the telescope's observations from the previous few days and circled the anomaly, famously writing *Wow!* in the margin in red pen. Interpretation constituted identifying the source of the signal — somewhere in the direction of the Sagittarius constellation — and the characteristics of the transmission (its intensity, frequency, bandwidth, etc.). But understanding remains elusive.

Dick also identifies a common "prediscovery phase," where "the true nature of an object, signal, or phenomenon goes unrecognized or unreported, or during which only theory indicates the phenomenon *should* exist." Theoretical physics suggested the existence of black holes long before they were discovered, but no mathematical formula anticipates the prevalence of life in the cosmos. Theory says *should;* imagination says *might*.

The Message

"As I imagine it," Carl Sagan said, "there will be a multilayered message. First there is a beacon, an announcement signal, something that says, *Pay attention. This is not some natural astronomical phenomenon. This is a signal from intelligent beings . . .* Then, the next layer is one that says, *This message is directed specifically to you guys on Earth. It isn't directed to anybody else.* And the third part of the message is the real content, which is a very complex set of data in a new language which is also explained."

It's here that we come to the crux, to the heart of things: to imagining how humans and aliens might meet. They move from *out there* to *right here,* or we transport ourselves to them. The future is nearer, imminent. And everything implicit in imagining aliens — the reflections and inversions, the comparisons to humans and life on Earth, the great relief of breaking through the confines of n=1 — is made explicit. In stories of contact we imagine not just what aliens might be like, but what humanity would be like once we met them.

Carl Sagan's novel, *Contact,* then, which he was describing above to Studs Turkel, is the heart of the heart. It's a 370-or-so-page answer, literally or in spirit, to every question we can ask about how finding alien intelligence might go. Yes, there's conflict and strife — acts of terrorism, government obstruction, frustration and loss and death — but at its core the story promises an inviting cosmos. A door opening to a galactic community. We're not only not alone but also welcomed. This hope is central to the idealistic origins of SETI, to Sagan's motivations as a scientist and communicator. It also makes it especially weird that the novel ends with its heroine finding proof that God is real, but we'll get to that.

That heroine is Eleanor Arroway (with one of my favorite names in all fiction), a SETI researcher struggling against skepticism and disdain for her commitment to the possibility of finding extraterrestrial life. (Both the book and Robert Zemeckis's film adaptation feature early scenes dramatizing the struggles of securing funding, perhaps the work's greatest fidelity to the facts of the scientific endeavor.) But then, Ellie's observatory receives a signal coming from the star Vega. It's a string of prime numbers, indisputably unnatural. Beneath the beacon of the prime numbers, the signal is decoded to be a rebroadcast of Hitler's 1936 Olympics address — the first Earth transmission made at a high-enough frequency to slip through the atmosphere. (Sending Hitler back to us turns out to mean *Hi, we got your transmission,* but it's certainly alarming at first.) Buried beneath that video is what comes to be called the Message, along with a primer at allows the Message to be read. As Ellie's ETI mentor believes, "They would try to make it easy, because if they wanted to communicate with dummies they would have to make allowances for dummies." The Message is instructions for building a Machine (also capital *M*). Its function is unknown, but humanity unites, if falteringly, to construct it.[46]

Some warn that it may be a Trojan horse doomsday device; others worry we're too far behind the Machine's designers for them to want anything to do with us, so their goals must be more nefarious than conversation. The technology is too advanced for its human contractors to understand, but just within their ability to fabricate. Step by step the Message prescribes the formulation of new materials and the construction of a massive Machine with space for five . . . passengers? No one knows what to expect

46 Well, two of it — one in Wyoming, one in the USSR; the Soviet Machine falls behind schedule, though, and when the Wyoming Machine is sabotaged, a benevolent, eccentric billionaire who has been funding the US project reveals a third site, in Japan, where copies of Machine components were being studied, where a new Machine can be built.

when the Machine launches on New Yea[r's] Eve, 1999. Will it take the five people with[in] it somewhere, or open a door for aliens t[o] come to us? To outside observers, the Machine spins for twenty minutes and then stops; but the travelers onboard — including Ellie — experience a much longer trip, through a series of wormholes to the center of the galaxy, where the senders of the Message await.

Nowadays, SETI researchers offer far less fantastical possibilities, reframing their search not as one for messages or signals but for technosignatures — like Jason Wright's Dyson spheres, or a radio signal leaking out from an alien transmitter with no expectation of galactic eavesdrop. But Sagan and his generation of the first SETI pioneers were looking for a capital *M* Message in their nonfictional lives, too. In a 1980 episode of *Cosmos,* Sagan mused, "Perhaps they'd send a compilation of the knowledge of a million worlds, the Encyclopedia Galactica." It would be part manual for advancement, and part census report: "We would choose some nearby province of the galaxy, a region that's fairly well-explored, and then slowly leaf through the worlds." Maybe, as in *Contact* (and *Star Trek: First Contact*), the aliens are waiting for proof that we're technologically ady, or maybe they're just broadcasting the encyclopedia — or Machine blueprints — nonstop, across the sky, to be received by anyone capable.

Sagan thought we were ready, but also, we were desperate. Whether it would be technological advice or a community census, the *Encyclopedia Galactica* seemed to be something humanity needed. SETI as a scientific discipline was born in the aftermath of World War II, at the dawn of the Cold War. Humanity was newly possessed of the power to truly destroy itself, and the detonation of two atomic bombs in war had not taken our fingers off the button. A signal from an advanced alien civilization would be proof that advanced civilizations could exist, could survive their own technological power and thrive. And maybe, Sagan and his colleagues hoped, whoever was out there would know that a nascent technological power needed help, in the form of technology or advice or something we couldn't even imagine.

In *Contact,* the Machine and the journey its passengers take are not the life raft. The life raft is the Machine's construction on Earth. The signal's discovery is major news, and religious leaders grapple with it. New sects and cults pop up around the world. But even before the Message is deciphered or the Machine is built, Sagan writes, "In a world gingerly experimenting with major divestitures of nuclear weapons and their delivery systems, the Message was taken by whole populations as a reason for hope . . . For decades, young people had tried not to think too carefully about tomorrow. Now there might be a benign future after all." The Message or the *Encyclopedia Galactica* is icing — it would be enough just to know someone is out there.

This is the logic behind Adam Frank's project from the last chapter, imagining paths through the Anthropocene for extra-terrestrial civilizations. He hopes his models create hope by presenting the possibility for us to course-correct climate change. Jill Tarter, the SETI scientist on whom Ellie Arroway is sometimes said to be based, believes SETI offers hope, too, that the power is in the search, not the discovery. She told me, "Our ability to do these searches has the effect of holding up a mirror to a human and saying, *Look in that mirror. Compared to something else out there, you are all the same. We humans are all the same.*" It's a bit of just finding a new Other against whom humanity can, conceptually, unite — hopefully in a more peaceful process than in, say, *Independence Day* — but Tarter believes that recognizing our unity as humans is crucial. "We have to figure out how to get along and to take better care of the planet, or else our future is short." Whether our looming fear is nuclear winter, climate change, or some other civilizational end, we hope for something or someone to shock us out of our inaction.

Ann Druyan calls this thing that SETI might give us "human self-esteem." Druyan was married to Sagan from 1981 until his death in 1996, and among their many collaborations,[47] she's the uncredited coauthor of *Contact*.[48] And it's that ephemeral

47 Druyan was the creative director for the Voyager Golden Record — more on that in a bit — cowriter of the original *Cosmos* series, and creator, producer, and writer of the 2014 *Cosmos* reboot/sequel.

48 Though Druyan was the one with fiction-writing experience, she was told, "No one wants to buy a book by Carl Sagan and his wife," but she was assured that if *Contact* was a success — which it was — she could be credited on their second novel. That promise was never fulfilled before Sagan's death. But she had previously published a novel in 1977 and is the credited coauthor of four books with Sagan, and in addition to being a producer of

enlightenment that's at the heart of the novel, more meaningful in the end than technological advances or cosmic journeys.

Ellie and the four other passengers are shuttled on a tour of the galaxy through what seems to them like a subway network of wormholes. They see Vega, the source of the Message, where a planet-sized sphere covered in radio telescopes orbits the star. They visit a host of stellar wonders — a binary pair of stars so close they touch, a brown dwarf that's not quite a star or a planet, a star flickering with changes in its intensity — until they are brought to a massive space station at the center of the galaxy. Its surface is covered in portals, docking sites of various shapes and sizes for, Ellie realizes, Machines conveying beings of all shapes and sizes, too. "The vision of a populated Galaxy, of a universe spilling over with life and intelligence, made her want to cry for joy." As the human Machine docks with the station, she thinks, "What a vindication for the human species, invited here at last! There's hope for us."

There, the aliens give Ellie and her companions almost everything they could hope for, save proof of their story to bring back to

---

the film of *Contact* is credited alongside Sagan for the story.

arth. Each passenger meets an alien Caretaker in the illusory form of a loved one. (Druyan told me this was because Sagan's number-one rule for writing an alien was never to show them, never to collapse the possibilities of imagination into anything you could clearly see.) For Ellie, this is her deceased father. She asks if they've passed the test, but he tells her that's not what this is about. "Don't think of us as some interstellar sheriff gunning down outlaw civilizations. Think of us more as the Office of the Galactic Census. We collect information." Ellie wants a taste of that information about the cosmos, the aliens, everything she doesn't know. But though the alien tells her he knew humanity needed help, what he really offers is a mirror. It's what Sagan offers the reader, too, hoping that some fraction of what real contact might do to humanity could be communicated through fiction.

"Last night, we looked inside you. All five of you." The alien gaze is loving, but it sees everything. "Love is very important. You're an interesting mix." He tells her, "I think it's amazing that you've done as well as you have. You've got hardly any theory of social organization, astonishingly backward economic systems, no grasp of the machinery of historical prediction, and very little

knowledge about yourselves." None of th
comes as much surprise, but it's still so reas
suring to be so seen and understood.

"You humans have a certain talent for adaptability — at least in the short term." But, he concedes, "You can see that, after a while, the civilizations with only short-term perspectives just aren't around. They work out their destinies also."

It's not a grim prediction so much as a warning. And the Caretaker does offer Ellie inspiration as well, with a view of the vast work ancient civilizations are capable of across the universe. He tells her of a collective of many galaxies engineering the cosmos.[49] "You mustn't think of the universe as a wilderness. It hasn't been that for billions of years," he says. "Think of it more as . . . cultivated." But even these gardeners and architects are not the most powerful, or most ancient, beings in the universe. The wormhole network that brought Ellie to the station was something that they found. Ellie asks, astonished, "There was a Galaxy-wide civilization that picked up and left without leaving a trace — except for the stations?"

49 This carries echoes of the later stages of cosmic enlightenment imagined by Stapledon in *Star Maker*, too.

Basically, yes, he tells her. And in other galaxies, too.

There always needs to be someone older or more powerful, it seems. Some mystery. The Star Maker behind the curtain. The tunnel-builders who, billions of years ago, ran off to no-one-knows-where, if *where* even applies. We're left with artifacts and clues, scraps of messages and wonder.

Ellie asks him to tell her about his myths and religions. "What fills you with awe? Or are those who make the numinous unable to feel it?" (She's been talking a lot with a handsome preacher.) And so he tells her about *pi*. And this is where, it seems to me, we're being told that God exists.

The Caretaker tells Ellie there's a message in *pi*. That when the infinitely long, nonrepeating decimal is calculated out far enough, "the randomly varying digits disappear, and for an unbelievably long time there's nothing but ones and zeros." A message in binary, undeciphered, woven into the very fabric of the universe. And when Ellie gets back to Earth and her New Mexico observatory, now funded by the government as long as she keeps the truth of her journey secret,[50]

50 The lack of hard evidence of their journey fuels belief that the whole thing was a hoax,

she turns the decryption algorithms th[illegible] she had once used on cosmic radio signa[illegible] to calculate out the digits of *pi* — and scan[illegible] for patterns, for a signal in the noise. On the very last page of the novel, her computer finds an anomaly in *pi* when expressed in base-11, a sequence of ones and zeros that, when printed out on a grid, reveals a perfect circle. "The galaxy was made on purpose, the circle said."

At first thought, I find it *extraordinarily strange* that *Contact,* a book about the power of scientific inquiry as a source of awe and self-discovery, would end with proof that God is real. Not the God of Jesus or Abraham or anyone else, but some creator, some Star Maker who embedded a message in math. Religion is a fraught theme in *Contact:* the power of Evangelicalism over American politics, the stigma against atheists (in the movie version of *Contact,* Ellie is initially denied her seat in the Machine because she is cornered into conceding that she does not believe in God), the tension between science and religion. But Ellie's journey is, indeed, one of faith — her SETI research is

---

which is the easier narrative for the government to promote while they look for more evidence.

motivated by a belief in something bigger, beyond Earth; upon her return, she has no proof beyond experience of the journey. In the movie, Ellie speaks even more directly in echoes of religion: "I had an experience. I can't prove it, I can't even explain it, but everything that I know as a human being, everything that I am tells me that it was real. I was given something wonderful, something that changed me forever, a vision of the universe that tells us undeniably how tiny and insignificant and how rare and precious we all are. A vision that tells us that we belong to something that is greater than ourselves."

But maybe the two aren't as distinct as I would think — faith in the numinous by way of alien contact, proof of the intentional creation of the universe. Druyan told me, "It's the laws of the universe as a kind of holy, sacred thing . . . Not punitive, not judgmental, not telling you what to eat or who to love. But the idea that the laws of this universe are knowable . . . There is something sacred in discovering these laws."

If God is in the fabric of the universe, then the Caretaker who comes to Ellie in the form of her father is something like an angel. His presence and the galactic works he speaks of may as well be gospel, and though she knows he's not her father, she still seeks his opinion, or something more like judgment, a verdict on humanity rather than on herself.

He offers her something else often found in religion: a broader context within which to understand the Earth. By expanding our sense of the scope of the world, science fiction (and fantasy) "have the same central function as myth and theology," writes Ryan Calvey in his doctoral thesis, *Transcendent Outsiders, Alien Gods, and Aspiring Humans*. And anthropologist John Traphagan points out, "It is no accident that SETI arose in a cultural context heavily shaped by Christianity and its inherent assumptions about the existence of a higher being." In *Contact*, the Caretaker offers Ellie not just knowledge and information but a benevolent attention focused on humanity, offering us a small nudge in the direction of peaceful survival. Even without a worldview with God, we want to be able to see ourselves through his eyes. For context, for insight, but also to know we're okay.

Even with our decades of technology and centuries of civilization, we are children in the gaze of these beings. But there's something reassuring about that, the same way I still want my mom when I'm sick. If we're children, then our mistakes are just the messy path of learning; if we're children, the grown-ups can still come and help. We don't want this violent, greedy, suffering version of humanity to be our final form. Transcendent outsiders give us hope and, hopefully, guidance. But even just knowing they are out there — and that they are reaching toward us — could be enough to change the world.

Even after the Machine seems to have failed, the American president[51] says that "the greatest good granted us by the Machine . . . was the spirit it had brought to Earth," that it showed humanity "we were all fellow passengers on a perilous journey in space and in time." Ellie is so frustrated that the Caretaker sent her back with no proof of her journey, but his strategy of (as the movie puts it) "small moves" seems to be working. We learn, "Many editorialists commented that this pause [after the Machine's supposed failure] was welcome." Both the flood of new science and new ideas brought to Earth by the Message needed time to be made sense of. Maybe it was for the best that true contact seemed to have eluded us, they think. "Sociologists and some educators

51 In the book, fictional and female, replaced by a Forrest Gump–style collection of Bill Clinton newsreels in the film.

claimed that the mere existence of extraterrestrial intelligences more advanced than we would require several generations to be properly assimilated."

The Search

SETI as a scientific field had three births: a 1959 paper in *Nature*, a 1960 search for extraterrestrial signals conducted by Frank Drake, and a 1961 conference attended by fewer than a dozen scientists. Why the confluence at this moment? In some ways, it was the first time we really had the technology. The development of radar during World War II facilitated the emergence of radio astronomy in the decade following. But the way we imagined space was changing rapidly, too. Sputnik launched in 1957, Yuri Gagarin went into orbit in 1961. The cosmos was opening up to us, vastness coming within reach. At the same time, imaginations were being primed by science fiction — *Forbidden Planet, Tales of Tomorrow, The Blob, Invaders From Mars*. The fictional aliens of the 1950s didn't need to be plausible to open up the possibility that real extraterrestrials might be out there, too. What John Traphagan calls "the seeds of a new imaginary" had been planted. And one way they bloomed was as SETI.

For its first few decades, SETI was less a scientific practice than a call to action. Though the premise was solid, it still seemed pretty fringe. Research was sporadic and piecemeal through the 1980s, hampered by shortcomings in funding and technology.[52] And just as SETI in the US was about to get some real momentum going, and solid footing with NASA funding, the plug was pulled in 1993. Senator Richard Bryan issued, on that occasion, a press release saying, "This hopefully will be the end of Martian hunting season at the taxpayer's expense." (Shame about the dozen-odd subsequent Mars missions, I guess.) Today, NASA is venturing into funding SETI research again, but most funding comes from private organizations like the SETI Institute and, more recently, individual donors like billionaire tech investor Yuri Milner.

The generation of SETI researchers to follow Sagan and Frank Drake has worked to move the field into a more agnostic stance. Not in terms of whether God left a message inside *pi,* but in terms of what they're looking

52 The USSR also had a SETI program, differently funded and a bit more cohesive than in the US. But otherwise, no other countries in the twentieth century really got into it.

for. SETI scientists used to talk about beacons and messages; now they talk about radio leakage and technosignatures. They used to stir up the public's imagination with speculation about the *Encyclopedia Galactica* and blueprints for Machines; now they turn their imaginations from whoever might be out there toward what it would mean to us, here, to know that someone was out there, once or now or long ago, no matter how little about that someone might be known.

Part of the shift in framing is from searching for intelligence to searching for technology — artifacts and evidence, with no assumptions about who made it or why. "We have this terrible problem," Jill Tarter told me, "we can't define life. And we can't define intelligence, right? We can simply, in our case, look for evidence that something has modified its environment, in ways that we can sense remotely over vast distances."

Jason Wright, whom we met in the last chapter searching for Dyson spheres and other evidence of massive engineering projects, is attempting to ground speculative science in more rigorous, rational frameworks. His approach may ultimately make the research more palatable to NASA, which hasn't funded SETI in thirty years. But even with those shifts, SETI can still seem like a quest rather than a series of scientific experiments. I made the mistake of suggesting to astrophysicist Andrew Siemeon, director of the Berkeley SETI Research Center, that his work was not hypothesis-driven. That is, I imagined that he was pointing a radio telescope to a corner of the sky and hoping to capture something — either he would or he wouldn't, but there wasn't a hypothesis being tested, right? Wrong, very wrong. "Our research seeks to quantify the distribution of advanced life in the universe," he told me, when I visited his lab in Berkeley. With every survey, a section of the sky is scanned on a given range of frequencies; with every null result, that's one more set of stars we know isn't transmitting toward us in that range. It's not a binary between detection and no useful knowledge, either we get a signal or the search was a waste, but a gradual delineation of what we know — the slivers of sky we've investigated, small surveys accumulating to, so far, confirmations of silence.

Sixty years old is still pretty young for a scientific field. Siemeon compared it to alchemy — not as a magical process but as a scientific discipline, before it transformed itself into proper chemistry. "We know there's a question that's a good question. And we know there's something interesting going on. But . . . we're trying to answer the question with whatever we have."

What we had in the 1960s was radio technology, hot off a boom of development. But just a few years later came the laser. While radio waves traverse long distances well because of their laconic, long wavelengths, laser can do the same with intense concentration. Charles Townes, who helped develop the laser, even wrote a *Nature* paper himself that essentially says "*I made this amazing thing, and this would be the best way to communicate across interstellar distances.* He knew that in 1961 . . . and I think he's right. That's how you do it."

That's astronomer Shelley Wright,[53] who's building an observatory to conduct optical SETI. While we've had radio telescopes since the 1960s, the ability to receive SETI signals in the optical spectrum is new. And it's commensurate with today's technology, not that of the sixties. Wright told me that today's lasers could be seen by telescopes like ours thousands of light-years away, and they would dramatically outshine the sun. And even as barely advanced as humans are, we're leaking less and less radio

53 No relation to fellow SETI researcher Jason Wright.

into the cosmos, verging on "radio-quiet." More advanced aliens might use radio even less. So Wright sees a vast, and unexplored, span of the electromagnetic spectrum to be observed.

But for all her enthusiasm for lasers and Townes's insights, Wright recognizes that these passions are hers, not extraterrestrials'. "It doesn't have to be a laser, it could be anything we didn't think of." There's speculation about aliens communicating in gravitational waves or with neutrinos or other unimaginable media, but we've barely scratched the surface with our radio search, and optical SETI is only just getting started. Why not search for what we know how to search for? We have no way of knowing how aliens might use their technology, but we can at least start with the simplest uses of our own.

Today, SETI takes an objective approach, questioning core assumptions of the field while avoiding any discussion of potential alien motives. Take Sagan's confidence in math and science as literally universal. In *Cosmos,* he promised that science would be "a cosmic Rosetta stone," or even more than that, a common language. But Jason Wright and others aren't so sure. Our math may be just one language among many for making sense of the universe. Concepts like prime numbers or infinity or counting may be neither universal nor fundamental — if not arbitrary, then intrinsically derived from human physiology or culture.

But even if we got a signal that was entirely indecipherable, it would at least be recognizable as not natural. The universe wasn't content to give me a midwriting scare with the Venus phosphine news — I would love to find alien life, I would not love for it to make me have to revise my entire book — and so, in December of 2020, word came that Siemeon's lab had found . . . something. There was a signal coming from the direction of Proxima Centauri, the closest star to the sun. It was a strikingly narrow signal, the equivalent of a dial tone. It was clear from the start that it didn't contain information. But it was also clear that, as far as human knowledge about the origin of radio signals goes, this one was technological.

By April 2021, though, analysis showed that the signal was, in the words of researcher Sofia Sheikh, "a pathological example of [radio] interference." At least from my view of the internet, this news gained much less traction than the initial leak of the signal's existence. The *Guardian* had broken the story, in December, before the Berkeley researchers were ready to go public, but most reporting hewed to Siemeon and Sheikh's careful line: *We don't know that it came from Proxima Centauri, just from that direction; it will take months of analysis to figure out the source; we know it's technological, but we assume that means it's from Earth until we fully exhaust those possibilities.* It did take months, and in that time, without further news or tantalizations, the public's attention wandered away.

The SETI researchers I spoke to are mostly dispassionate, too, about what would happen if they did confirm a signal was extraterrestrial and technological. Siemeon said (when I interviewed him before the news of the Proxima Centauri signal), "I've been accused of being arrogant in this particular area, but I just don't think about it a lot. I think it would be good — I can't imagine any way that it would be bad. So I don't worry too much about how it's all going to play out." Jason Wright told me, "My personal best guess is that even if we were to find a communicative transmission, we would not be able to decode it." He likened it to giving Thomas Edison a modern cable modem and seeing if he could access all the information on the internet. And John Traphagan proposes, "These imaginaries will occupy the general public for a while and scholars perhaps for a long time. And then, given that the distances will be great and the time lag between 'hi' and 'how are ya' will be vast, we will return to our usual business of greed, poverty, hunger, and war."

If we found a signal, it's entirely possible it would come from a planet hundreds of light-years away — Proxima Centauri as a source was an almost too-lucky fluke, just a bit over four light-years distant — and its senders could be long gone: extinct, evolved into a new civilization, departed from that planet to seek a new home. It could be like finding a letter from Queen Elizabeth — the first one — and trying to answer her. And then hoping anyone remembered, centuries into our own future, to listen for a response.

I imagine it could be something like Solaristics, the fictional field of study that develops around the inscrutable alien entity in *Solaris*. Dispatches from a distant planet could become a through line for humanity, like the cyclical recurrences of Halley's Comet — *Oh, sometime next year we should receive a message.* Scholars would brush up on the last message, get their receivers ready. Maybe we'd have settled into a set schedule, or maybe for a year or two a fleet of radio telescopes would be focused, waiting. And then the message comes. Maybe it takes years or decades to be decoded and for a response to be agreed to. Maybe human civilization reorganizes itself around this effort. Or maybe the work hums along in the background while most people go on with their lives.

But what if luck were on our side, and a signal came from closer by? The conversation could happen faster — in years instead of centuries — but imagine it's a bit further in the future, and human technology's a bit more advanced. Could we really resist the temptation to go meet the messenger face-to-face?

The Journey

Carl Sagan imagined a signal that announced itself with prime numbers — universal, incontrovertible, the perfect beacon for an astronomer to find. But when anthropologist Mary Doria Russell found herself inspired to write a novel of first contact, she thought, "Personally, I wouldn't walk across the street to meet someone who thought prime numbers were enticing. But music? . . . I could imagine being pulled across space by music." And so in her novel, *The Sparrow,* that is what humanity hears.

The music in *The Sparrow* isn't a message, either (at least not to humans). It's signal leakage, the sort of inadvertent transmission most of today's SETI scientists think is what we'd be more likely to find. Instead of being crafted by benevolent, advanced aliens, everything about the signal seems shaped by serendipity — or perhaps divine hand. It turns out to come from the Alpha Centauri system, only a bit more than four light-years away. "They are too close not to go," one character says. "The music is too beautiful." And before the international community can figure out what to do about it, a far more nimble organization — well versed in first contact — takes advantage of every bit of kismet to consummate their mission. "The Jesuit scientists went to learn, not to proselytize. They went so that they might come to know and love God's other children . . . They meant no harm."

Russell's imagining is no less idealistic than Sagan's, but with no super-advanced Caretakers to guide humans through making contact, even the best intentions are not enough.

The planet Rakhat is in many ways hardly alien at all. A planet in a three-sun system, sure, but other than that — "The niches were all there. Air to fly through, water to swim in, soil to burrow under . . . The principles were the same: form follows function, reach high for sunlight, strut your stuff to attract a mate." And the people who live there are strikingly familiar, too. They are bipedal and bilaterally symmetrical, they laugh and yawn and smile, they speak vocal languages that can be learned by humans. But when a book's prologue ends with *They meant no harm* — when a book's inspiration is in part the five hundredth anniversary of Columbus's journey to the Americas — you know things won't end well.

When the human crew lands and makes contact, they quickly realize that the alien people they've met are not the Singers whose transmission they received. The aliens' technology is rudimentary, and furthermore their language is not the language of the songs. That's fine, there's no reason to presume a culturally homogeneous planet. But the anthropologist-author who imagined music drawing humans across the stars is actually a *paleo*anthropologist whose most significant academic work shaped our understanding of Neanderthals as human. Rakhat is home to two sentient species.

Even after meeting the second species, the Jana'ata, more advanced and quick-witted than the docile Runa the humans first met, it takes the visitors terribly long to understand what they've found: a world with two sentient species, a predator and their prey.

The balance of this world is delicate. Inspired by predator–prey relationships on Earth, Russell imagines that the Jana'ata population would be much smaller than the Runa. She points out in interviews that humans have domesticated our own prey plenty of times, breeding for docility and meat, and we've bred dogs for intelligence and to do various jobs; the Jana'ata just did this all at once, and with another species of people. But though it is a balanced system, the humans cannot abide it. Even when they see the full picture — the system is so carefully calibrated that no one goes hungry, no one is unemployed — there remains the fact that people are being bred to be eaten.

The humans have no policies of noninterference, no Prime Directive as *Star Trek* has to stop them from meddling with an unprepared world. Just their hearts, full of good intentions. The humans have a small technological advantage over the Jana'ata, perhaps a few decades ahead, but that matters far less than their very presence as visitors — the destabilization, the ability to interfere. So many stories of first contact, written by white American or European authors, replay the horrors of colonization by putting America or Europe into the position of subjugation by colonizing aliens, karmic comeuppance for our culture's past sins, imagining that the worst horror would be to be subjected to the atrocities to which we once subjected other humans. But Russell goes in another direction, imagining the best of intentions, and then the heartbreak and horror of inflicting great suffering nonetheless.

No one is left more heartbroken and horrified than Emilio Sandoz, priest and linguist and the only member of the human crew who survives to return home. Upon first meeting the Runa and finding his Earthly linguistic techniques will work here, "He felt as though he were a prism, gathering up God's love like white light and scattering it in all directions." Unlike stories like *Contact,* where humanity is blessed by an encounter with a superior being, here transcendence comes from the love of finding other beings in the universe and recognizing in them the kinship that Sandoz understands as being God's children. This is not just receiving a signal or identifying an artifact, but truly *making contact*. Emilio is "smiling and in love with God and all His works." That can be a Christian God or the Star Maker or the universe itself. The sequel to *The Sparrow* ends with an echo of *Contact*'s message in *pi*, an echo of *Star Trek*'s humanoid progenitor: a message encoded in the combined DNA of humans, Jana'ata, and Runa — not in geometry, this time, but music. Ellie Arroway says, "They should've sent a poet," but Russell suggests that they should've sent a priest.

Sandoz's face is said to evoke, depending on the light and the moment, sometimes a conquistador and sometimes a Taino person — it's his Puerto Rican heritage and his narrative lineage, too, this uneasy alignment with European conquest. The humans in *The Sparrow* may come to Rakhat with no intent to conquer or profit — sincerely, they seek to know God's other children — but Russell still wrote with that historical inspiration. *Somebody ought to write a story that would put modern, intelligent, well-educated, well-meaning people into the same position of radical ignorance as Columbus and his men. Let's just see how well we'd do!*

Stories of historical first contact between cultures on Earth are, in all their paths and permutations, models for imagining contact with aliens — for fiction writers like Russell and scientists alike. When Stephen Hawking famously cautioned against humanity making ourselves known to aliens lest they come to kill us, he did it through the lens of this analogy: "If aliens visit us, the outcome would be much as when Columbus landed in America, which didn't turn out well for the Native Americans."

But there are so many stories of contact, and so many ways to read them. Hawking's comment aside, SETI scientists and analysts use analogies drawn from Earth's history to support all sorts of conclusions. You can cite examples to claim that contact will be violent or benign, that technology will disrupt civilizations, seed religions, or lift a developing culture to sudden advancement. But the most famous, most disastrous contacts loom over it all. Genocide, epidemics, slavery, destruction.

We hope that those analogies won't hold and that others might supplant them in our hopeful imaginings. Since contact is so much more likely to happen across great distances and thus across time as well, another common analogy is the transmission of ancient Greek knowledge to medieval Europe, the unearthing and piecing together, the subsequent Renaissance. But maybe that translation was too easy to parallel grappling with a message from alien minds. Is it more accurate to think of it like modern

English-speaking scholars trying to parse Mayan inscriptions? What about the scant sense we can make of the Lascaux caves?

We stretch and stretch, more alien, more distant. All of these analogies try to bridge an impossible gap. But sometimes we need to tell these stories, insufficient as they may be. Scientists, at least, need to justify funding seemingly speculative projects. Analogies help them say, *We've trodden a path like this before. Our work could change the world, and here's how.*

Hawking's not alone in framing the conquest and genocide of the Americas as a triumph of technology; whether that's his bias as a physicist or just how he learned the story, who can say. But scholars now attribute the conquest's expedience much more to the decimation of the New World's inhabitants in the face of Old World disease, novel pathogens as tools of genocide themselves. "[T]he stories we've been raised on," Kathryn Denning writes, "stories about technology and power and politics and how civilizations meet, and what happens next, require our critical attention instead of our unthinking allegiance."

We can't really think our way entirely out of analogy. These are the shapes we know. After all, as Denning writes, "[E]very time we tell a story about what will happen in the event of a detection, or contact, we retell the story of contact here on Earth." But first contact between cultures is more than an analogy, it's a history still reverberating in the lives of millions of people. And some of those people write stories about aliens, too.

In the hands of Western, white authors, stories of alien contact often become a re-writing of past stories of first contact, along the line of Russell's inspiration for *The Sparrow*. What if it were now, what if we were well-intentioned — what if "we" were the conquered? Science fiction scholar Rachel Heywood Ferreira sees this conceit resonating not just in stories of conflict but in the guardrails set up in fictional worlds to avoid recapitulating these awful outcomes. She writes, "The prominence of the Prime Directive in the *Star Trek* universe and the prevalence of similar injunctions against contact and interference with more primitive, planet-bound civilizations are surely rooted to a great extent in such regrets." And anthropologist William Lempert writes that "a central purpose of such prime directives throughout history was not to protect the vulnerable, but rather to morally legitimize colonial enterprises," as if the regret were anticipated.

But Heywood Ferreira is mainly interested in exploring what stories of alien contact are like when the author doesn't identify with the colonizer at all. "For those writing in a postcolonial reality," she writes, "the consequences of contact/conquest/colonization are especially immediate, woven into the fabric of both everyday reality and cultural identity." As Lempert puts it, "History does not simply rhyme, it reverberates."

Looking at first contact fiction by Latin American authors, Heywood Ferreira writes that these stories might better be called *second contact,* so present is historical first contact in their authors' minds. You can feel that shadow in African sci-fi as well. "Where tales of first contact are concerned," Heywood Ferreira writes, "there is a pervasive sense of déjà vu . . . that 'this has happened here before!' "

The writers feel that déjà vu, and often their characters do, also. In Nnedi Okorafor's *Lagoon,* in which aliens appear in the waters outside Lagos, Nigeria, there's a sense of surprise that aliens would begin their visitation — or is it conquest? Is there a difference? — in Africa, since the dominant shape of visitation stories, told from dominant nations, centers on the global superpowers. One character thinks, "He didn't know if he believed in aliens or not. He'd never considered the question. If there were aliens, they certainly wouldn't come to Nigeria. Or maybe they would."[54] But there's also a sense of preparedness in the déjà vu. Another character thinks, "He was alive, and worse things had happened. He chuckled. This wasn't the first invasion of Nigeria, after all."

In Tade Thompson's *Rosewater,* also set in Nigeria, a more subtle but also more nefarious invasion is met with a similar lack of surprise. "We have more experience than any Western country in dealing with first contact," one character says. "What do you think we experienced when your people carved up Africa at the Berlin Conference?

54 The aliens' choice of Lagos also facilitates their plan — and opens up the narrative space for exploration. While first contact is familiar in Lagos, alien contact is less so. "If they'd landed in New York, Tokyo, or London, the governments of these places would have quickly swooped in to hide, isolate, and study the aliens. Here in Lagos, there was no such order." The lack of rails for the story to run on, and lack of coordinated government response, leaves more space for human nature to play out in all its forms and directions.

You arrived with a different intelligence, a different civilization, and you raped us. But we're still here." Thompson is Nigerian British; he said in an interview that setting the novel "in Nigeria was a no-brainer. Being a former colony gives a country a better perspective on alien conquest." In a review of the last two books in the *Rosewater* trilogy, critic Jessica FitzPatrick points out that while the book's Nigerians "successfully formed pockets of resistance or adapted to living with the invaders . . . imperialist countries like Great Britain and the United States had less productive positioning [against the alien visitors] and were quickly destroyed or driven into an isolationist retreat by the threat of alien invasion."

Thompson didn't set out to analogize African history but, he said, "I have a rage. Anybody whose ancestors have had to go through that, we'll have that rage. And there is no way that that rage isn't going to bleed into my work." He initially found his way into the story by asking himself why aliens would visit Earth. "What makes the energy and personnel expenditure of interstellar travel worth it? . . . I felt that the standard invasion trope of a big-ass mothership announcing arrival and shooting everything to dust didn't make sense except perhaps from a colonialist perspective." For Thompson, the aliens' sophistication would play out more subtly, "steering rather than dominating."[55]

That steering, he realized as he was writing it, was the aliens' form of neocolonialism, the colonization of culture and ideology. Thompson said, "I think that I was on the second draft before I realized that pretty much the entire novel could be seen as a metaphor for neocolonialism. That all of it could be."

He deliberately used analogy to understand the experience of his characters, imagining what it would feel like to be living on the African coast in the 1600s — "You've been kicking around with a canoe and all of a sudden you see a ship like a tea clipper turns up with sails and everything . . . That is what alien contact must feel like, if there were any aliens. If a flying saucer arrives that is the same kind of fear." But the analogy isn't an intellectual exercise. Thompson is imagining himself into his own history, to carry that thread through to an imagined future. "It is also not the case that SFF [science fiction and fantasy] is exclusively about

55 Dominating "with mind-blowing weaponry which," he adds, "has an annoying knack of stimulating rebellion."

the future. It is about the present, and it is about reclaiming narratives of the past."

Where the Meaning Lives

Some fictional aliens are so advanced they seem — or actually are — psychic. Perhaps they've been studying humans from afar to prepare for an invasion, or they come with universal translators, or they can flat out read our thoughts. But in other stories, the challenge of learning an alien language makes those shortcuts by comparison feel like getting your nutrition from a pill. Sometimes there's pleasure in having to chew.

Sue Burke, the author of *Semiosis,* also works as a translator, and her love for that vocation runs through her essay about Ted Chiang's "Story of Your Life." Perhaps a bit blinded by enthusiasm to Louise's personal suffering, she calls the story "the adventure I want to have: aliens arrive at Earth. Someone has to learn to talk to them. Let me do it! Me! Me, me!" She isn't just eager to speak to aliens — whether it's Chiang's heptapods or anyone else who might be out there — she's giddy with imagining their linguistic possibilities, the ways biology or culture might shape (or be shaped by) a language's bones. She writes:

What might be required in an alien language? Perhaps an entity with a distributed or inherited intelligence would need to specify the internal origin of the utterance. A creature that communicates with light might have a grammar that branches like a decision tree in infinite paths, each step specifying a choice made and not made. The language of far-ranging diplomatic robots might have been designed to be easily understood by incorporating mathematics into language to represent relationships among concepts.

In *The Sparrow,* Mary Doria Russell gives ample space to these imaginings. Her retort to *Contact*'s *They should've sent a poet* might not have been *priest* but *linguist,* Emilio Sandoz being conveniently both. It doesn't hurt that the Runa the humans meet are traders, culturally and perhaps genetically predisposed to openness with strangers, and that they train their children to be interpreters. In the humans' first encounter with them, a young Runa girl, Askama, comes forward. Emilio says hello, and she echoes his English. Then she greets him in Ruanja, and he does his best to repeat the greeting. Emilio touches his chest and says his name; Askama repeats

"Meelo." The Runa and Emilio both kn this dance.

At first, Sandoz proceeds by naming things He shows Askama a flower, and she says, *"Si zhao."* He shows her two flowers — with the flourish of a magic trick — and she says, *"Sa zhay,"* which he files away as possible evidence of plural formation.

Later, he learns the peculiarities of how Ruanja conceptualizes nouns. He says:

> "One day, Askama showed me a very pretty glass flask and used the word *azhawasi*. My first guess was that the word *azhawasi* was more or less equivalent to *jar* or *container* or *bottle*. But one can never be certain, so one tests. I pointed to the side of the flask and asked if this was *azhawasi*. No. That had no name. So I pointed to the rim and again, that was not *azhawasi* and had no name."

But it turns out that the Runa name functions, not objects. *Azhawasi* is the space enclosed by the unnamed, in Ruanja, flask. "Similarly," Emilio says, "there is a word for the space we would call a room but no words for wall or for ceiling or floor."

It feels, to an English-speaker, like a suitably alien way of understanding the world.

hen I see a cup, I see a cup, not the space nclosed by it. I can't imagine seeing a wall and not having a word for the surface. Ruanja has different declensions for nouns based on whether they are seen or not. (A person who is not with you counts as unseen just as an abstract concept like love does.) Russell has said that she based her alien languages on human ones, presumably with sounds strange enough to the English reader's ear. She used traveler's phrase books to constrain herself to non-Romance, non-Slavic sounds: Nepali for the Jana'ata language, K'San and Quechua for Ruanja. When I asked her over email, she said that to the best of her recollection (having been asked out of the blue about a writing decision she made thirty years ago) she didn't borrow the spatial/nonspatial distinction but probably invented it herself.[56]

56 She followed up to tell me that, on discussion, her husband had pointed out that she sees the world in strange categories herself: "Like in the fridge, there's a door compartment for Things You Spread on Baked Goods (butter, cream cheese, peanut butter, jellies, and jams). I have another place in the fridge for Things You Squeeze: mustard and ketchup bottles, squirt tubes of garlic, ginger, tomato paste,

These feel like unnatural rules to my bra and to the brains of the humans of the nove — but they're still principles that can be understood and learned. Emilio learns and teaches Ruanja just like he would handle any language on Earth. Russell imagines many forms of convergence between life on Earth and life on Rakhat: plants and animals, predators and prey, laughter and body language and biology. But the fact that Rakhat's languages can be learned by humans at all, as if they were just particularly strange Earthly tongues, would be its own kind of convergence, momentous for our understanding of aliens and our understanding of the idea of language, both.

In a 1983 interview, linguist Noam Chomsky said, "If a Martian landed from outer space and spoke a language that violated universal grammar, we simply would not be able to

---

etc. There's a kitchen drawer for Things That Can Kill You, which includes knives, a heavy meat pounder and a corkscrew. In the basement work room, I have a box of Things That Connect Other Things, which means glues, rolls of tape, bungee straps, balls of string . . . So I'm pretty sure now that I made up the grammar you asked about."

ırn that language the way that we learn human language like English or Swahili." ınstead, he said, the study of Martian would be more like that of physics, "where it takes generation after generation of labor to gain new understanding and to make significant progress." Again, the emergence of a field like Solaristics comes to mind, of academic debate rather than stark discovery. And it depends on the answer to the question: Is human language representative of all language or just how language happens to work here on Earth?

Chomsky's idea of universal grammar is prominent in linguistics but by no means universally accepted. The concept has evolved over the decades, but the essence is that there is something in human brains that allows us to learn (human) language. Chomsky wasn't inspired by some complete cataloging of human languages and emergent universal qualities but by observation of the best language-learners on Earth: young children. While adult learners find different languages to be of varying levels of difficulty — due to similarities with their native language, complexity of rules, prevalence of irregular formations, and other factors — children all tend to acquire language within the same time frame. And they do it quickly, without ever really needing to taught.

I've found myself feeling so silly sometimes defining a word to my toddler. "A garage is like a house for cars when they're not driving." He doesn't need that. He wakes up one day meeting all surprises with "What the heck!" deployed perfectly correctly. Young children don't seem to learn languages so much as absorb them. Just as stem cells are undifferentiated and can develop into any specialized form, linguist Andrea Moro calls a child's mind a *stem mind*. But this isn't a blankness, at least according to Chomsky. It's a preloaded map.

The human brain has evolved to take immersion — not instruction, just ambient language — and turn that into fluency. In the broadest sense, this hard-wiring is universal grammar. In a narrower sense . . . it gets complicated. Linguist Alec Marantz, who studied with Chomsky, told me, "The tension in the field at the moment is between the understanding of universal grammar as just the name for whatever it is in the brains of humans that makes them able to learn language . . . versus claims about the specifics of that." Some researchers identify elements of human language that they say are universal — patterns of grammar, foundational rules

— and call those universal grammar, but these claims are hard to pin down, as their categorizations are necessarily subjective. A deeper definition of universal grammar gets at not just how humans learn language but what makes human language unique among communication systems we know on Earth.

Beneath the linear order of a spoken, signed, or written sentence lurks the two-dimensional structure — a tree or a map — of syntax. Words placed distant from one another can still be in close meaningful connection, because linear order doesn't determine (or reflect) meaning; structural relationships and hierarchy do. Just in that sentence: words . . . can . . . be. It doesn't matter that *distant from one another* and *still* intervened, your brain knows that *words* is what *can be,* and even that's a relatively straightforward example.

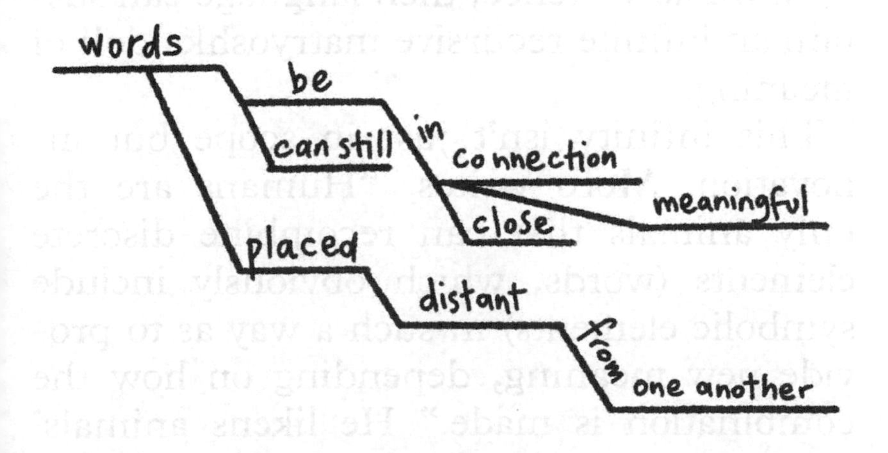

This is the magic on which all human language runs, and the structure of syntax allows for the quality that Chomsky identifies as essential to language: discrete infinity. With a finite number of words, humans are able to create infinite sentences and meanings, because syntax allows us to nest phrases within phrases, to build detours and extensions and additions with our words, until our sentences are like frankensteined McMansions — they may be strange and unwieldy, their kind unknown before this moment, but they're still grammatically sound and, most imporantly, you can understand them. Moro offers a tidy illustration of this infinite power. Take two sentences: *Mary left,* and *This made John sad*. In that pairing, a single word, *This,* holds in its reference an entire other sentence. If a single word — in its own sentence, of course — can stand in for a whole sentence, then language can sustain an infinite recursive matryoshka doll of meaning.

This infinity isn't just in scope but innovation. Moro writes, "Humans are the only animals that can recombine discrete elements (words, which obviously include symbolic elements) in such a way as to provide new meaning, depending on how the combination is made." He likens animals' vocabularies to "dictionaries of sentences," fixed and discrete. This is, of course, a difficult assertion to make conclusively — who knows what mysteries of therolinguistics we haven't yet cracked. But animals can learn new vocabulary, like a dog learning *Sit,* and songbirds can combine their song elements in new and original ways, but no new meaning is created in the process. It seems very much that only humans can utter new sentences, and with them, we can imagine new ideas.

But is this the nature of human languages, or the essence of language itself? An alien might communicate without language, as Earthly animals do, but what richness could their communications or cultures find while limited to preset sentences, no innovation through recombination toward infinity? If it's impossible to have language at all without these qualities, then what Chomsky calls universal grammar would not be a principle of human language but of language itself, and those Martians, if they have language, could have one that humans can learn.

But maybe the question isn't *Discrete infinity or not?* but *Do they share our universal grammar?* What if there are weirder languages in the cosmos than are dreamt of in our linguistics? Maybe an alien syntax is 3D in structure, spoken without reduction in dimensions at all. Maybe the medium isn't sound or gesture but scent or something imperceptible to us. Maybe categories we think of as essential, like verb and noun, like speaker and word, are merely local conventions, hopelessly parochial even as we yearn to converse across the stars.

There are plenty of alien languages humans can learn in science fiction — especially onscreen, where human actors have to speak the lines. The forerunner was Klingon (though, I think you could argue Tolkien's Elvish is a kind of alien language), which was created by a linguist and reverse-engineered to fit a few Klingon phrases that had made their way casually on-screen. But it was fleshed out enough for devoted fans to become fluent speakers (and for my father to give part of his toast at my wedding in its guttural barks and bellows).

David Peterson faced a similar challenge as Klingon's creators on his most famous project: creating the Dothraki language for *Game of Thrones*. From the handful of words and phrases coloring George R. R. Martin's books, he built a complete grammar and vocabulary. From words in the books like *khal, rakh, haj,* and *dosh,* he built other consonant-vowel-consonant words, like *vezh, chek,* and *jith,* and longer words that fit the given patterns. From a few sentences in the books he constructed an entire word order, grammar, and lexicon. And all within the known culture of the people who would speak it — Dothraki has words for *kale* and *carrot* but not *computer* or *please*.

Peterson makes his living inventing languages for TV and movies — for sci-fi, fantasy, and a surprising number of witches. I asked him which was the most *alien* alien language he'd invented, but he said, "Most of the time, those languages are not very alien. And they're not very alien because the aliens themselves are not very alien." Constrained by the realities of special-effects budgets and actors' bodies, these aliens are, as far as language is concerned, pretty human. "They're bipedal humanoids, they breed, they mate in the same way, they raise their young, they eat, they drink, they excrete, they live a fixed amount of time, they have a similar working memory." Part of the rigor Peterson applies to his languages is that they suit the creatures who will speak it. And so if the aliens are not very alien, the language can't be either. In those cases, he said, "It was more like creating a human language for a strange culture that doesn't exist on our planet."

The closest Peterson's ever gotten to creating a truly alien language was a project just for himself, too strange to be useful on a film set. "I tried to imagine a language that would treat nouns and verbs as identical. Such that when they were arranged, you could easily swap one out for the other — it wouldn't be seen as doing anything weirder than swapping out *I kicked the ball* versus *I kicked the wall*." This was over a decade ago, and he eventually abandoned the project because, he said, he just wasn't good enough yet to pull it off. But it was his most meaningful attempt to "get at something that was truly nonhuman, that truly defied what we do with our languages."

Defying what we do with human languages is linguist Andrea Moro's specialty. Moro set out to test the bounds of language by constructing an "impossible language," one that contradicted rules that he suspected were universal in the human language-using brain. He points out, in his book, *Impossible Languages,* "If we were biologists, we would not hesitate to claim that there are impossible animals: an animal that produces more energy than it absorbs, for example, or an animal capable of indefinite growth. We could make such a claim because all organisms are constrained by physical laws, like entropy or gravity." And so, could we not imagine a language that similarly violated whatever laws constrain language?

Moro drilled down to the most essential nature of language: that word order in a sentence doesn't dictate meaning. Syntax isn't linear but a two-dimensional map (as in the example diagrammed above), where rules are hierarchical, not linear. The linear sentence is a representation of that 2D structure, reduced by one dimension, as a circle is a reduction of a sphere. (A language might require an element to come first or last in a sentence, but never, for example, third, where the placement would be determined by the spot in line rather than its place in the hierarchy.) Moro tested this idea by teaching German-speakers a boiled-down version of another language — Italian or Japanese — with some alterations. Some rules operated as they normally do, but others were based on sequence in a sentence. For example, they were taught that in a negative sentence, the word *no* needed to come fourth.

It turned out that while the subjects didn't find the "impossible" rules harder to learn than those of natural syntax, and they didn't notice anything fishy, their brains knew the difference. When subjects were learning natural, hierarchical rules, the language center of the brain, Broca's area, activated; when they studied and practiced the linear rules, Broca's area quieted down.

This may seem a bit tautological: the language center of the brain was active when learning language. But this study shows the clear line between possible language and not, and that's the line between hierarchical rules and those based on word order. Essentially, only certain communication systems register in our brains as language. Thus, it's possible only certain communication systems can be learned through our natural, inborn aptitude, but this is my own stretch.

Syntax allows us to create infinite meanings constrained by a finite word list — and linear word order. Time and the human vocal apparatus limit us to one word after another, but our brains somehow track the syntax humming beneath the surface. "Words," Moro writes, "come in sequences: this is perhaps the only incontrovertible fact about human language." But it is quite possibly a human limitation.

In "Story of Your Life," Louise recognizes that the heptapods have two distinct languages: she calls their spoken language Heptapod A and their writing system Heptapod B. Louise observes of the latter, "If I wasn't trying to decipher it, the writing looked like fanciful praying mantids drawn in a cursive style, all clinging to each other to form an Escheresque lattice, each slightly different in its stance." She calls the biggest sentences "sometimes eye-watering, sometimes hypnotic." And, it turns out, becoming fluent in this nonlinear language gives her access to the heptapods' nonlinear experience of time.

In the movie adaptation, Heptapod B looks more like ink blots arrayed on a circle, trailing fronds like dye seeping through water. It's beautiful but, David Peterson pointed out to me, not actually nonlinear: the line is just drawn in a circle. Peterson did introduce me to a visual language that's truly nonlinear, called the Elephant's Memory. Developed by author and designer Timothée Ingen-Housz beginning in 1993, the Elephant's Memory consists of about a hundred and fifty logograms, pictures that stand in for elements of meaning, which are arrayed into sentences with rules that determine their form and placement. (Being nonlinear, sentences like *I made the house burn* and *The house burned because of me* are identical.) Peterson pointed out a sentence to me: *Seeing elephants shot by men makes me cry.* He said, "You'll notice it's not in a circle. But at the same time, it's nonlinear, in that you could start anywhere, and the whole thing expresses the idea."

TIMOTHÉE INGEN-HOUSZ

Nonlinearity is easy enough in writing, but the nature of time and vocal cords constrains our spoken speech to one word at a time. Systems like sign language do open up possibilities of simultaneity. In American Sign Language, for example, ideas are conveyed simultaneously with hand gestures, facial expression, and bodily cues, among other elements. And of course, we even layer meaning onto spoken language with inflection and body language. But for language itself, as Moro writes, "Words come in sequences."

What if, fiction must then ask, *they didn't?* Then you end up with a new kind of impossible language, a speech that layers meaning as music layers harmony.

China Miéville's *Embassytown* conjures one such possibility, which turns out to be "impossible" for humans in more than one way. The Hosts are an alien race, so-called because their planet, Arieka, hosts a human settlement — the titular town. The Hosts, or Ariekei, have a vaguely insectoid anatomy — scuttling legs, wing-limbs, a coral constellation of eyestalks. But they lack what many other "exot," or alien, species have in this vast inhabited cosmos: what Miéville calls a shared "conceptual model" for the world.[57]

This alienness manifests in Ariekeine language. They speak, simultaneously and always in chorus, from two mouths — "inextricable by the chance coevolution of a vocalizing ingestion mouth," like we have, "and what was once probably a specialized organ of alarm" — though even that, in this universe, isn't so weird. "The Hosts aren't

57 There are other kinds of exots in *Embassytown* that are more familiar, mindwise, to humans, including one race that's immortal, barring accidents. (They are also rarely born.)

the only polyvocal exots. Apparently there are races who emit two, three or countless sounds simultaneously, to talk. The Hosts . . . are comparatively simple." But Ariekei speech is almost not language, in that it doesn't signify, it . . . is. "Where to us each word means something, to the Hosts, each is an opening. A door, through which the thought of that referent, the thought itself that reached for that word, can be seen." (Here we walk the fine line between the incomprehensible and magic.)

The humans who first met the Ariekei quickly made sense of their language but found that no matter how perfectly they programmed computers to speak the two-voiced words, the Ariekei didn't understand, didn't even seem to understand that language was being spoken. It was only when a pair of linguists, in luck and frustration, together shouted the two halves of a greeting word, that an Ariekene recognized that humans were trying to communicate at all. "The Ariekes turned to us. It spoke. We didn't need our [translation software] to make sense of what it said. It asked us who we were. It asked what we were, and what we had said. It had not understood us, but it had known there was something to understand . . . [I]t knew that we had tried to speak."

As one human character explains, if he were to program his language software with a human language, a human would understand it. "If I do the same with a word in [Ariekene] Language, and play it to an Ariekes, I understand it, but to them it means nothing, because it's only sound, and that's not where the meaning lives. It needs a mind behind it."

Here, the essential languageness of language isn't syntax but representation. When speech is a conduit to meaning, even a word like *that* is impossible, because it could be unspecified and can mean *this* or *the other,* a violation of the essential requirement that speech be grounded in truth. The closest the Ariekei come is their use of simile. Similes let them express new ideas, as they do for us. But because every utterance in Language must be a truth, the Ariekei orchestrate their similes, breaking and mending a rock to be able to say *It's like the rock which was split and fixed.* Once humans came to Arieka, they became part of the palette; as a young girl, the book's main character was asked to perform a simile, so that the Ariekei could add to their vocabulary *It's like the human girl who in pain ate what was given her in an old room built for eating in which eating had not happened for a time.* You know, like *that*. Does it help to know it will eventually be shortened to "Like the girl who ate what was given her"?

It may be hubris to imagine that, with work, we could understand an alien language. With live speakers, sure. The experimentation and inference that Emilio Sandoz uses in *The Sparrow* or Louise uses in *Arrival* are the core practice of field linguistics. But in the much likelier (still possibly very unlikely) case of an intercepted message, we would have no interaction, just static text. There are still dozens of Earth languages (or writing systems) that remain untranslated, human mysteries uncracked. A deliberate message might come with a primer — in *Contact* it uses basic arithmetic to teach logical relationships, which then allows symbols and words to be defined. But how can we hope to communicate across light-years if we can't even understand ourselves across a few millennia on Earth?

The Most Ambitious Expectations

When we think about communicating with aliens, it's not just translating a message we've *received*. We're also devising and sending messages, too, out to the cosmos for whoever might find them. We've seeded the haystack with a few needles ourselves.

In 1974 we beamed the Arecibo message

in the direction of M13, a star cluster about 22,000 light-years away. The 1,679 binary digits of the message form a little tapestry image representing some basic scientific information and a stick-figure man. But the message was meant more as a demonstration of human technological capability than any true greeting to aliens. The Arecibo dish was new, a powerful telescope and transmitter. M13 was chosen not because it was a good place to start a conversation, but because its apparent size fit nicely within the beam of the transmission.

But the most meaningful messages we've sent to the cosmos haven't been beamed at all. They've been carefully constructed, shot into space on rockets, and hurled through the solar system with such force that they'll just keep going.

I have a part of that message tattooed on my leg. It's often mistaken for a firework or sparkler, a long stem leading from my anklebone up my calf. The only people who know what it means are those who are already familiar with the symbol. But it was originally designed for viewers who would know even less what to make of it, if they even had the senses to process it.

The symbol is called the pulsar map, a representation of our solar system's place

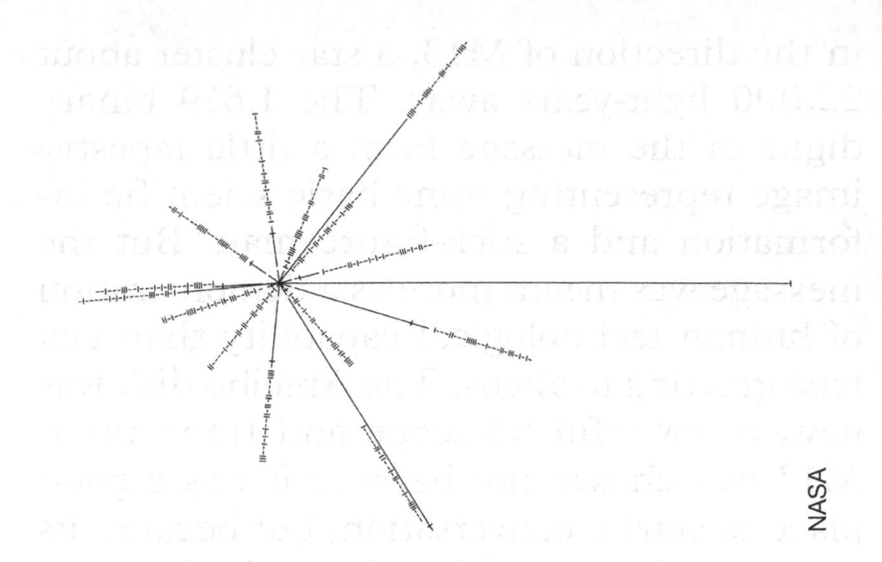

in the galaxy, cross-referenced against the galactic center and fourteen pulsars, each represented by one ray of the starburst design. (A pulsar is the remnant of a massive supernova, now an impossibly dense neutron star, rotating at incredibly high speed. It shoots a powerful beam of radiation that, thanks to its spin, sweeps a lighthouse signal through the cosmos.) Here, their distance from Earth is shown in the length of their ray, the frequency of their pulse is shown by a dashed line, and their orientation in the galactic plane — the galaxy being disk-shaped — is indicated by a little tick mark somewhere along the ray.[58]

58 I did have to simplify the schematic to make it a viable tattoo, so if I am found floating

The idea is that if an intelligent alien comes across this map, floating somewhere out in space, they could figure all of that out.

The pulsar map is onboard four of the five human-made objects to achieve the escape velocity necessary to leave the solar system. It was part of the design of the plaques affixed to the 1972 and 1973 Pioneer probes, along with a little schematic of the probes' journey past the planets and a Vitruvian-style man and woman. The pulsar map is also on the cover of the Golden Record, the much more robust messages-in-a-bottle with which the Voyager probes left Earth in 1977. (The fifth escape-velocity object, the New Horizons probe that photographed Pluto and is currently exploring the outer solar system, was adorned with a less cohesive epistolary payload, a collection of Earthly objects including an American flag, two state quarters,[59] and a piece of hardware from a failed early-2000s private-spaceship initiative.)

The Golden Record is the richest message humanity has sent into the cosmos, devised by a committee led by Carl Sagan over the

somewhere out in space, I will be impossible to return to sender.

59 For Maryland and Florida, the sites of the craft's construction and launch.

course of a few months. Its cover contains clues to its origins — the pulsar map — and instructions for playing the record, converting the etched grooves on its surface into images and sound. The images include scientific diagrams and photographs, of the Arecibo telescope, of a mother breastfeeding, of a woman standing in front of a supermarket produce display, eating a grape she likely hasn't yet paid for. The sounds include over two dozen works of music (including Bach, Javanese gamelan, and Chuck Berry), greetings in fifty-five languages, and recordings of thunder, wind, trains, automobiles, heartbeats, and whalesong. There's also an hour-long recording of Ann Druyan's brainwaves, taken as she meditated on the history of humanity, its current struggles, and "a personal statement of what it was like to fall in love."

Druyan and Sagan fell in love during the six months of the record's creation, but that wasn't the only romance. She speaks as lovingly of the Golden Record as she does of Sagan. She told me, "One of the reasons I have such an emotional attachment to Voyager is it is one of those occasions for human self-esteem. It hurt no one. And yet in every way it exceeds the most ambitious expectations of its makers. It is that great place where our brains, our souls, our hearts and our music . . . combined in one work of art and science." The Voyagers will be circumnavigating the Milky Way for the next billion to five billion years, up to ten orbits of the galactic center (where Sagan and Druyan imagined in *Contact* the Caretaker's welcome station to sit). "And to be able to do this twenty years after Sputnik?" Druyan said. "That is a learning curve. That really gives me hope."

Druyan knows — as everyone working on the record knew at the time — that the odds of either spacecraft being found out in space are infinitesimally small. It was in so many ways a project for humans — a mirror to hold up to ourselves to say *This is who we are*. Music and whalesong and the laughter of children, the brainwaves of a woman newly in love, so in love that it colored even her meditation on the great suffering of a planet. If the Arecibo message was a demonstration of human technological ability, this is a demonstration of something more complex, multifaceted — an attempt to demonstrate an essence of humanity itself.

But infinitesimally small odds are still something. Druyan told me, "With the probability of *more than zero,* over five billion years —" she leaned toward me and smiled as she held up her hand "— *two* spacecraft" (that's double the odds) "you start to get into that neighborhood of *Well, you know, it's possible.*" She leaned back with a small smile and a shrug.

She told me she's imagined that possibility playing out a trillion times — the extraterrestrial finding a Voyager, decoding its schematic, setting the record to play. And then she wonders: What could the sound essay mean to them?

For decipherment purposes, the greetings in fifty-five languages are a nightmare — instead of a cleanly organized Rosetta stone, they're a mishmash, all over the place, each speaker saying whatever they were moved to say. You can't learn a thing about their meaning by putting them next to each other, and you'd give yourself a headache to try. The photographs, two-dimensional representations of a three-dimensional world, depend not just on humanlike visual systems but on human conventions that are as culturally anchored as they are biologically.

But, Druyan told me, she feels optimistic that her brain waves, the record of her hour-long meditation, could be deciphered and understood by aliens who found it. "I would love for that meditation to be something that everyone could see. Because it was so joyful. And just so new. True, complete, total love, suffusing even the narrative about the planet and the life on it and our plight, and how messed up we are. All of that, and how much we're capable of feeling for each other. That would be amazing."

As for imagining who might find the record, Druyan told me, "The eons it takes for nature and the environment and mutations to sculpt life is so splendid that when I'm yearning to see the extraterrestrial, I think it's going to be the way I felt when I first laid eyes on my two babies. That sense of, *Of course, that's what you look like. You're so beautiful. That makes perfect sense. Could I have imagined you into being? No.* And that's how I think perhaps we'll feel when we actually finally do, in some shape or form, make contact. That sense of *Oh, yes, of course.* But not that we have the wherewithal to imagine it."

Today, the Voyager probes are more than twelve billion miles from Earth, more than a hundred times the distance between the Earth and the sun. There's no clear line between the solar system and interstellar space, but if you put that boundary at the farthest reaches of the solar wind, they've crossed it.[60] They still have 20,000 years, though,

60 If you put it at the extent of the sun's gravitational influence, or the Oort cloud's halo of

before they'll leave the sun's gravitational influence and 30,000 years before they'll be near another star (the red dwarf Ross 248).

But beyond its boost to human self-esteem, I think there's another way the Golden Record is more for us than for any aliens. We're the only technological beings that we know of. So the likeliest interstellar travelers to come across the Voyager probes could very well be far-future humans. Even if the probe doesn't get rewired by alien AI into sentience like V'ger in *Star Trek: The Motion Picture,*[61] it could, given 10,000 years, become alien to us — because we could become alien to ourselves.

Ten thousand years isn't an arbitrary measure of time. I bring it up because it's exactly the gulf across which we are trying to communicate warnings for nuclear waste disposal sites, the contents of which will be dangerously radioactive for a hundred centuries. How do we make a warning that will stay intelligible for just as long?

A common way to emphasize the scope of this challenge is to take a walk through the

matter around the sun, the probes have quite a ways to go yet.

61 An imaginary Voyager 6, to be fair.

history of the English language. Middle English, spoken until about six hundred years ago, is barely comprehensible to a contemporary English-speaker; Old English, from a thousand years ago, isn't comprehensible at all. It's a matter of entropy, in a way, information subject as anything else is to the second law of thermodynamics — not decaying out of order but morphing and shifting, small changes adding up over the centuries until you're somewhere entirely new. And that's with a relatively contiguous culture. Imagine a disaster, or a social break: humanity survives, perhaps fully rebuilds, and encounters a mystery in the desert. You need these future humans to not only understand the warning but also believe it. All the promised curses in the world didn't stop nineteenth-century archaeologists from broaching the sealed doors of the pharaohs' graves, after all.

In 1993, the U.S. Department of Energy convened a commission of anthropologists, linguists, semioticians, and other experts to figure out how to make an immortal message. Every option they proposed feels like a conceptual art project, but maybe that's not so surprising. When you discount any carrier of meaning that's tied to our current culture — a placard that says Keep Out, or the segmented-circle sign for radioactivity — what you're left with is the visceral, nonverbal evocation of feelings that are, we hope, universal for humanity. In this case, the panel hoped for the evocation of fear, awe, or disgust. *What is here is dangerous and repulsive to us. This message is a warning about danger,* as well as *This place is not a place of honor . . . no highly esteemed deed is commemorated here . . . nothing valued is here.*

The panelists ended up covering their bases with redundancy: Rosetta stone–style plaques in half a dozen languages, imposing granite columns surrounding the site, warnings in words and image and mood. Three rooms will hold identical information about the project: two hidden, buried and walled with granite, and one at the center of the array of columns. To let in natural light, it would have no roof, I think because we have no idea if in 10,000 years humans will still have anything like flashlights.

Discarded ideas brainstormed by the commission include a field of massive stone spikes, set at random angles like thorns, to convey terror and inhospitableness. A report commissioned for another project along similar lines, seeking to develop 10,000-year warnings for a site at Yucca Mountain, had floated a generational game of telephone, breaking the millennia into 300 generations, and tasking every third generation with adapting their great-grandparents' message into new, contemporary language, so that it might not be irredeemably alien to their own great-grandchildren. Also in that report, also dismissed: seeding a 10,000-year folklore, a mythic false trail that would use superstition rather than scientific knowledge to urge people to shun the site. It would be overseen by an "atomic priesthood" in the know, the keepers of the radioactive truth. (The proven ineffectiveness of warnings at the pharaohs' graves made this one even easier to dismiss.)

Words, maps, scientific notation, landscaping, abstract sculpture. Would they be more effective than line drawings and schematics? Take a look at the Pioneer plaque, the Golden Record's simpler cousin. You're a human, this was made by other humans, but what parts of it do you understand?

The two people, for sure, because you've been raised in a symbolic culture that represents three-dimensional forms with line drawings. Maybe the lineup of the planets of the solar system, whispering to yourself *My Very Excellent Mother* or your preferred mnemonic, the incantation a cultural legacy

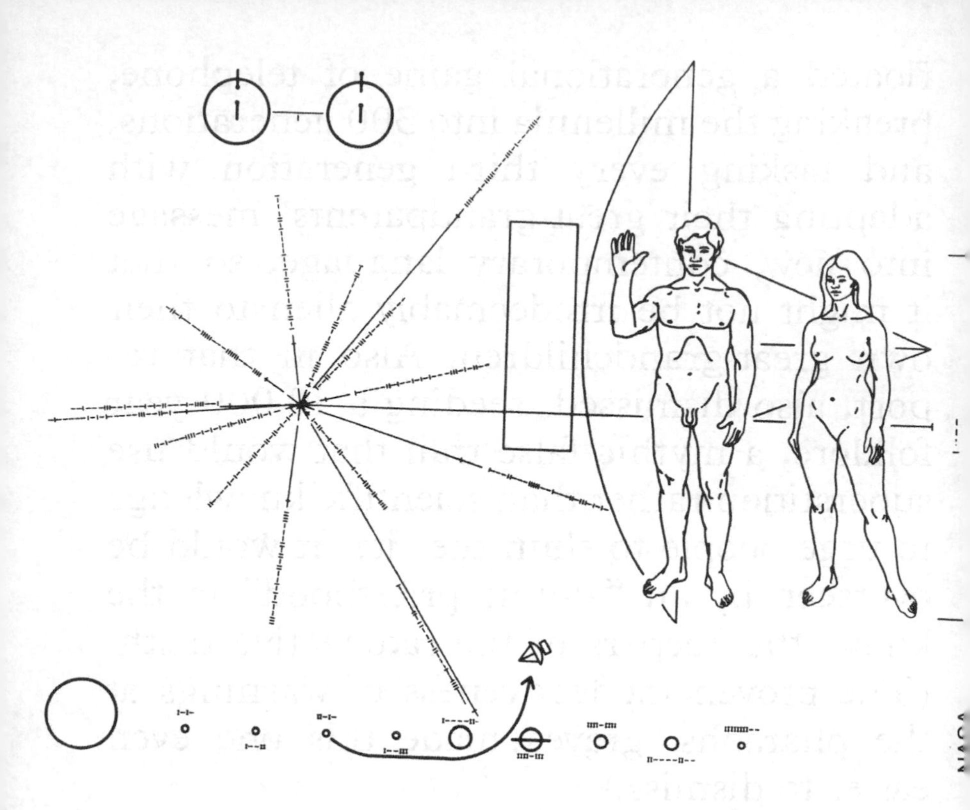

of its own. You understand the directional meaning of an arrow, a line with two smaller lines falling off its end, because your ancestors were hunters who used arrow-tipped spears.

We've sent these messages into space with little practical hope — they are needles in a cosmos of a billion haystacks, so unlikely to be found that why should we even worry if their message can be deciphered? The Voyager probes are about as big as a mid-size car, after all. Floating in interstellar space. (Though maybe an alien ship built for interstellar travel, with powerful-enough sensors to find our drifting jalopy, would be so advanced that they could crack our incomprehensibly alien code. We hope, we hope.)

It's all wishful thinking — that aliens find the probe, that humanity lasts another 10,000 years. But for now, these messages offer us a chance to project a hopeful, idealized version of humanity. Welcoming, laughing, suffused with love — looking forward, looking outward, and not alone.

HOPEFUL MONSTERS

I've sidestepped, as much as possible, questions of odds in this book. I'm less interested in the guesstimations than I am in the possibilities — what might life be like elsewhere, what would that mean for life on Earth? But all of those imaginings are haunted by what I've come to think might be a wildly rare event: the emergence of complex cells.

The fear that complex cells might be rare came to me from the pretty pessimistic book *Rare Earth*. The 2000 bestseller leaned all the way into odds and probabilities, making the case that the emergence of complex life is incredibly unlikely — Earths as planets are probably not rare, the authors supposed and research has since borne out. But plant-and-animal-covered Earths, intelligent-species-inhabited Earths? Those, the authors argue, may indeed be one-in-a-cosmos.

In the decades since the book's publication, many points of the authors' argument have been thoroughly argued against. They say we needed Jupiter to protect us from rogue comets; it now looks like Jupiter does some deflection but may also wing interlopers into the inner solar system just as often, rendering its effect roughly net neutral. They say the chemical building blocks of life are rare; by now, complex organic molecules have been found in meteorites and interstellar clouds. They say plate tectonics and a big moon have given us tides and the carbon cycle and Earth's stable rotations; I refer you back to Chapter 2 for just a hint of the ways we can imagine life around that.

But I continue to be plagued by what seems like one very unlikely event on the ancient Earth: the time one single-celled organism sucked up another and thus entered into a symbiotic relationship that would eventually give rise to all complex cells.

All multicellular life — and plenty of single-celled life, too, like amoebae and paramecia — has cells full of complex structures. Their genes are cloistered in a nucleus, and they have a host of internal organelles, the litany of whose names may flash you back to high-school biology: ribosomes, lysosomes, Golgi apparatus, endoplasmic reticulum (that last a beautiful set of words to rival *cellar door*). Ribosomes translate genes to proteins, lysosomes sequester enzymes from the rest of the cell, the Golgi apparatus packages proteins to shuttle them off to their destinations, and the endoplasmic reticulum folds and transports protein, among several functions.

The organisms with those cells are the eukaryotes. Prokaryotes, on the other hand, have no such complexity. These organisms, constituting two of the three base branches of the tree of life (bacteria and archaea, structurally similar but chemically distinct), have made plenty of chemical innovations since their origin a few billion years ago, but they remain, internally, soupy blobs.

Nick Lane, whom we met in Chapter 1 as an advocate for the energy-driven origin of life, writes:

> Speaking as a multicellular eukaryote, I might be biased, but I do not believe that bacteria will ever ascend the smooth ramp to sentience, or anywhere much beyond slime, here or anywhere else in the universe. No, the secret of complex life lies in the chimeric nature of the eukaryotic cell — a hopeful monster, born in an improbable merger 2,000 million years ago, an event still frozen in our innermost constitution and dominating our lives today.

In the late 1960s, evolutionary biologist Lynn Margulis proposed that our cells' complex structures came from what Lane calls "an orgy of cooperation, in which cells engaged with each other so closely that they even got inside each other." The scientific name for that orgy is *endosymbiosis,* where a mutually beneficial relationship between two cells leads to one moving in with the other — moving to live inside the other. It is now thought that most organelles evolved through other means, but the first organelle, and the trigger of all that complexity, did result from endosymbiosis. Eventually, the cells that were subsumed lost most of their autonomy. But their genomes, the relics of their independent origins, remain.

Inside your cells, one organelle is something of an interloper. Mitochondria, which turn oxygen and food into chemical energy your body can use, have only a handful of genes, just enough to make the proteins they use for their business. But those genes are not, in a sense, yours. Distinct from the genome in each cell's core, mitochondrial genes are passed down each generation in parallel, mother to child.

Billions of years ago, what would eventually become mitochondria was an independent organism. Lane and William Martin

believe a bacterium, which could metabolize oxygen, came to live inside an archaeon. In the process, all the possibilities for complex life were opened up. Afforded the output of their new internal engines, Lane and Martin's theory goes, cells now had energy to fund the evolution of other complex structures and, eventually, multicellularity. Martin has called the resulting eukaryotes "an archaeal genetic apparatus that survives with the help of bacterial energy metabolism." And that subsumed powerhouse seems to have been crucial, the key that unlocked complexity. "Natural selection," Lane writes, "operating on infinite populations of bacteria over infinite periods of time, should not give rise to large complex cells." It was just this one-off event. And they believe it happened only once.

Those mitochondrial genomes that tell us about these organelles' independent origins also tell us that the origin was shared, pointing back to a common ancestor. Just one. A single archaeon engulfed one bacterium, and their marriage (or something even more intimate) led to all the life you can see with your naked eye and plenty of smaller life besides.

Evolutionary biologist Mohamed Noor reminded me that chloroplasts, the organelles that conduct photosynthesis in plants, also have their own genomes, and so also seem to have arisen through endosymbiosis. But Lane and others argue that mitochondria came first and were a prerequisite. Complex life does not require chloroplasts; photosynthesis is something of a nice bonus. But no mitochondria? No complexity at all.

The timing of the origin of life on Earth suggests it was easy or probable. It happened just about as soon as conditions on the young planet allowed. But eukaryotes didn't emerge until about two billion years later. Two billion years of simple cells swimming around, evolving diverse chemical systems but nothing structurally interesting or new. The timing, plus the singularity of the event, starts to almost make it seem like luck. Martin calls it "unspeakably rare." Lane writes, "There is no innate or universal trajectory towards complex life. The universe is not pregnant with the idea of ourselves."

But as always, there's a flip side. In a paper titled "Eukaryogenesis, How Special Really?" philosopher of biology Austin Booth and evolutionary biologist W. Ford Doolittle answer their titular question with *Not very!* They suggest that the unique lineage of mitochondria we see today does not mean there were never others.[62] Perhaps archaea swallowing up bacteria for mutual benefit was all the rage two billion years ago, made probable by the environment or a certain evolutionary quirk. Lots of endosymbiotic pairs may have sprung up, perhaps across a variety of species. Ours is just the only one that survived . . . and gave rise to all multicellular life on the planet.

Booth and Doolittle also challenge what they call eukaryocentrism, this obsession with the emergence of our type of cells as a lightning-strike turning point in Earth's evolutionary history. And I do keep caveating: *multicellular life, all the life you can see with your naked eye.* Prokaryotes do amazing things, and I still keep referring to them as *slime*. But of course, we care about this single event because it's our history, a moment of rare luck that we carry with us now, humming away in our cells. Was it rare, was it not, does it only matter because it's the key to our existence here on Earth billions of years later? Would a cosmos full of planets inhabited only with sludge make us feel like we weren't alone?

62 Think of the Burgess Shale from Chapter 3, the myriad strange body plant life experimented with early on, though only a few survive to this day.

I don't like to think about whether-or-not. And I especially don't like to think about *this* whether-or-not, because it's tempting to lean quite heavily toward the *not*. But a weird thing happened when I was writing this book. I realized I'd sort of ceased to care.

The mitochondria were part of it. Not in their origin but their function, the vast complexity of chemistry that keeps us alive, evolved from humble beginnings in the churn of a deep-sea vent or coastal tides. The Burgess Shale was part of it, too, learning about the wild diversity of life that nature experimented with before, for whatever reason, life on Earth took this path. The dolphins and bats, with their incomprehensible minds. And the human researchers determined to find communion across the gulf.

When I interviewed astrobiologist Abel Méndez, I laughed in disbelief when he told me he doesn't care about finding life beyond Earth. ("Really??" my transcript says I said, and I can still hear my incredulous and slightly nervous laugh.) It seemed impossible, and a bit absurd. I mean, live your life, embrace your truth. But *I* was still full of longing for life beyond.

Yet the next phase of my research — working weirdly backward — was on the study of the origin of life. I talked to Lane and Martin as well as Sarah Walker and Lee Cronin and several other researchers digging into the questions of life's origin and life's essence — what is it that changes in matter when it becomes alive? Boundaries between this world and others began to fade. It wasn't life elsewhere that held the magic, anymore. It was that life existed anywhere, at all. We were just as improbable, and as important, as whatever we hoped might exist on worlds beyond.

I found myself, one morning, standing in my backyard while my dog trotted ahead to go to the bathroom at the edge of the woods. It was early spring, and I was surrounded by birdsong. The winter crows hadn't left yet, but a few grackles and sparrows had settled in, and one swooped in front of me on its way from tree to tree. And I found myself marveling. These alien creatures, able to eschew the ground for flight, alive to a cacophony of meaning in chirps and twitters that was entirely opaque to me. Their feathers and hollow bones and delicate, scaly legs. They were amazing! And incomprehensible. And real, right there. And I thought about Abel Méndez again and realized he was onto something. Or at least that I was onto the same thing.

In his book, *The Vital Question,* Nick Lane plays a poetic Ms. Frizzle and takes the reader on a Magic School Bus journey to the molecular scale. Now the size of a molecule of ATP, the cell's chemical currency, we're shuttled through a membrane's pore to the inside of a mitochondrion. Protons bubble up from the floor, and massive assemblages of molecules protrude from the walls. Lane describes the motion of ions and electrons zipping from station to station of these molecular machines, which tower above you. I picture the engine room of the *Titanic,* but a hundred times bigger. It's a city block of skyscraper-sized complexes of molecules, swinging and churning and flitting all to achieve one thing: to transfer protons across the membrane, so as to build the molecules of ATP that carry energy around the cell. It happens dozens of times a second, protons and electrons Rube Goldberg-ing down the line. Zoom out, and in the human body's forty trillion cells, in your quadrillion mitochondria, almost as many protons are pumped as there are stars known in the universe every single second. Turning breath and food into energy and life.

As many protons as there are known stars in the universe . . . Hold on to that for a moment. We'll come back to it.

This description showed me a chemical process vital to life, at the boundaries of biology, chemistry, and physics — the electrons that are moved down the line actually hop by quantum tunneling. Each electron's appearance and jump shifts the machinery just a hair, tweaking its posture here or there in response to the movement of charges. Lane writes, "Small changes in one place open cavernous channels elsewhere in the protein. Then another electron arrives, and the entire machine swings back to its former state." It makes life seem miraculous. It makes the invisible machinery of biology tangible and real. And it's just one of countless feats of biochemistry happening at every moment in every cell — every cell of yours, every cell on the planet.

This wasn't my first journey inside a mitochondrion. When I was about eight years old, right when I was starting to watch *Star Trek* with my dad, I also read Madeleine L'Engle's Time Quartet for the first time.

If *Star Trek*'s great dream is of an inhabited cosmos, *A Wrinkle in Time* hopes for an interconnected one. If *Star Trek* shows the possibility of diverse worlds, L'Engle wants us to see unity. Her books take us into the vastness of the cosmos and into microcosmic worlds as well.

In *A Wrinkle in Time,* young Meg Murry crosses time and space[63] to save her father from a malevolent power on an alien planet. There are aliens, yes — some eerily humanoid, some mythical, some incomprehensible — and similarly varied worlds. In the next book in the series, *A Wind in the Door,* Meg must save her brother's life from within his mitochondria.

L'Engle collapses and transcends physical scale, making a microscopic organelle as much a place as a planet is. Meg is shown the birth of a star, "so small that she could have reached out and caught it in her hand." Much like Lane does, L'Engle expands the infinitesimal until a little boy's body is a galaxy. But at the same time a newborn star can be held in a girl's hand.

One of Meg's somewhat magical companions, named Progo, tells her that his previous assignment was to learn the names of all the stars. She asks him how many there are.

"What difference does it make?" Progo says. "I know their *names*. I don't know how

63 Or, more accurately, she folds them so she can skip across.

many there are. It's their names that matter."

In this book, as in lots of fantasy novels (as this is maybe more a book of magic than of science), Naming is a power far beyond identification. It's a recognition of truth and an act of love. To Name a being is to know their essence, and once you see their heart, how can you not love them?

The numbers don't matter, the census or population. Just to know and imagine your way into another's world. To imagine their existence as vividly as you know your own. And to learn your own more deeply in the process — the alien stowaways who've made their home in your cells, the possible people on alien planets, the birds or bats in your own backyard.

The stars cannot be counted, but each one can be named.

ACKNOWLEDGMENTS

Six years ago, Lindsey Weber invited me to get coffee and talk about ideas I might have for an essay series on culture for *Medium*. This book would not exist if she had flinched when I said, "What about aliens? But through . . . a cultural lens???" and her wise edits on those pieces helped them become parts of this book, and made this book a possible dream.

A few years before that, this project was born in Patty O'Toole's research seminar at Columbia, when Patty told me to go broader than just the Golden Record. And I couldn't have written any of this or anything about space at all if Caleb Scharf hadn't let me, as a writing grad student, into his undergrad astrobiology class even though I couldn't do calculus. He even wrote me word problems to answer instead of the calculus problem sets. His ongoing generosity and support have given me the skills and confidence to do this work, and I try to live up to his model, as a teacher, mentor, and writer.

My agent, Katie Grim, has been a champion for this book and my work, and I'm so grateful she brought us to Hanover Square Press, where John Glynn has brought so much enthusiasm and insight to this book in his editing and Eden Railsback has been an indefatigable guide through the process. Thank you for believing in this weird chimera. Vanessa Wells' copyedits were incisive and felt like they came from a kindred spirit. Nibera Bernarda Conič's art on the cover makes this book feel like itself, and Sean Kapitan's art direction makes it a beautiful cover. I'm so grateful to the whole team whose work made this book more than words and ideas, including production designer Natasa Hatsios; Dayna Boyer, Rachel Haller and Randy Chan in marketing; Emer Flounders, Heather Connor, and especially Justine Sha in publicity; and Angela Hill and Grace Towery keeping the editorial wheels spinning.

This work was funded in part by a book grant from the Alfred P. Sloan Foundation, and I'm hugely grateful, especially to Ali Chunovic for guidance. And to Stuart Firestein, without whom I wouldn't have known about the grant or how to apply.

So many scientists, scholars, artists, and authors gave their time to tell me about their research, writing, and curiosities, and whether they're quoted here or not, each conversation made its mark on me and the book. Rebecca Charbonneau, Norman Johnson, Kamila Muchowska, Bridget Samuels, and Alex Teachey read early chapters and saved me from far too many careless or weighty mistakes. And Rachel Garner, fact-checker extraordinaire, made this book so much better and so much more correct, an invaluable collaborator. Any errors that remain are entirely my own.

Thank you to Jacob Shea and Jasha Klebe, composers of the *Planet Earth II* soundtrack, to which most of this book was written and revised. I'm not kidding. And to Chrissie for your transcripts, an invaluable resource.

In the years I spent figuring out how to write this book and then trying to write it well, Angela Chen, Meg Flaherty, Helena Fitzgerald, Kea Krause, and Becca Worby read early drafts and excerpts and helped me find my way. Various permutations of NeuWrite read various permutations of this project and especially helped me with Chapter 2. Carl Zimmer's visit to NeuWrite to talk about his own book triggered my ideas for the structure. Molly McArdle and

Jess Zimmerman, precious brilliant editorial minds, gave deep reads of the first draft.

Many friends fielded anxious texts and DMs about writing and publishing (and so much more, with so much love) over the last few years: Emily Adrian, Isaac Butler, Adrienne Celt, Nicole Chung, Alexis Coe, Rachel Fershleiser, Carl Erik Fisher, Josh Gondelman, Emily Hughes, Katy Kelleher, Swapna Krishna, Lucas Mann, Kate McKean, Arianna Rebolini, Helen Rosner. So many science journalists also offered guidance and support: Meghan Bartels, Kat Eschner, Lisa Grossman, Eva Holland, Shannon Stirone, Josh Sokol, Ed Yong, plus everyone in the best Slack. Tim Requarth and Adam Mann gave me invaluable direction and commiseration in writing about the origin of life. Meehan Crist kept me busy every autumn and kept me feeling like I could do this. In teaching me how to teach, Nicole Wallack, Sue Mendelsohn, Aaron Ritzenberg, and Glenn Michael Gordon taught me how to write this book (which they may notice is one big conversation essay).

Leslie Jamison showed me how to turn things inside out to crack them open (and, with Clarence Coo, gave me invaluable library access). Torie Bosch was the best boss a freelancer could ever ask for (and also came in clutch with library help). Kate and Jeremy Medow, thanks for being our CT family these last few years, and hopefully a few more. Ben, Buddy, Matt, James, I really fucking hope our D&D campaign is done by the time this book is out? I hope we didn't all die??? And to, as Siri reads it, "Breastfeeding Mother, Person Shrugging," Meg and Kea, writers and mothers and life support, I love you so much.

Allison Morris, you believe in me more than almost anyone in the world, and your love, encouragement, excitement, and generosity keep me going.

Thank you to my family for loving me and putting up with me: Marissa, Mom and Mike, Dad and Bonnie, Dawn and Tom, Mark and Sarah. Dawn, thank you for swooping in when we needed you. Bobie, thank you for showing me what it means to be an artist. Dad, thank you for Star Trek and the natural history museum. Mom, thank you for always supporting my weird art dreams, I don't take that for granted.

Tanner, how do I thank you for everything in the last few years that have been new parenthood, pandemic, interstate move, too many jobs to count, not to mention the whole book thing? We're in it together, and that's everything that matters.

And Miles — Coco — this book is for you: the whole universe, the future, every bit of love and hope and joy.

BIBLIOGRAPHY

Watchful Stars

Le Guin, Ursula K. *The Lathe of Heaven.* Diversion Books, 2014.

Chapter 1: Origins

"About Life Detection." 27 June 2022, https://astrobiology.nasa.gov/research/life-detection/about/

"All Good Things . . . part 1." *Star Trek: The Next Generation,* 7x25, 23 May 1994.

"The Chase." *Star Trek: The Next Generation,* 6x20, 26 Apr. 1993.

"The Devil in the Dark." *Star Trek,* 1x26, 9 Mar. 1967.

Astrobiology Magazine Staff. "Life's Working Definition: Does It Work?" NASA, accessed 28 June 2022, https://www.nasa.gov/vision/universe/starsgalaxies/life%27s_working_definition.html

Chen, Irene A. "The Emergence of Cells During the Origin of Life." *Science* 314, no. 5805 (2006): 1558–1559.

Cornils, Ingo. "The Martians Are Coming! War, Peace, Love, and Scientific Progress in H.G. Wells's 'The War of the Worlds' and Kurd Laßwitz's 'Auf Zwei Planeten.'" *Comparative Literature* 55, no. 1 (2003): 24–41.

Damer, Bruce. "David Deamer: Five Decades of Research on the Question of How Life Can Begin." *Life (Basel, Switzerland)* 9, no. 2 (2019): 36.

Davies, Paul C.W. et al. "Signatures of a shadow biosphere." *Astrobiology* 9, no. 2 (2009): 241–249.

Dick, Steven J. *Space, Time, and Aliens: Collected Works on Cosmos and Culture*. Springer Nature, 2020.

Gresham College. "Energy and Matter at the Origin of Life." YouTube, 28 May 2019, https://www.youtube.com/watch?v=vEZJdK5hhvo

Guthke, Karl. *The Last Frontier: Imagining Other Worlds, from the Copernican Revolution to Modern Science Fiction*. Cornell University Press, 2019.

Lane, Nick. *The Vital Question: Energy, Evolution, and the Origins of Complex Life*. W. W. Norton & Company, 2015.

Neveu, Marc et al. "The 'strong' RNA world hypothesis: fifty years old." *Astrobiology* 13, no. 4 (2013): 391–403.

Noor, Mohamed. *Live Long and Evolve: What Star Trek Can Teach Us about Evolution, Genetics, and Life on Other Worlds*. Princeton University Press, 2020.

Pressman, Abe et al. "The RNA World as a Model System to Study the Origin of Life." *Current Biology: CB* 25, no. 19 (2015): R953–963.

Requarth, Tim. "Our chemical Eden." *Aeon,* 11 Jan. 2016, https://aeon.co/essays/why-life-is-not-a-thing-but-a-restless-manner-of-being

Richardson, Sarah M., and Nicola J. Patron. "Synthia: Playing God in a Sandbox." *Microbiology Today Magazine,* 10 May 2016, https://microbiologysociety.org/publication/past-issues/what-is-life/article/synthia-playing-god-in-a-sandbox-what-is-life.html

Sagan, Carl. "Definitions of Life (Chapter 23) — The Nature of Life." *Cambridge Core,* 10 Nov. 2010.

Scharf, Caleb. *The Copernicus Complex: Our Cosmic Significance in a Universe of Planets and Probabilities*. Scientific American / Farrar, Straus and Giroux, 2014.

Scharf, Caleb. "Until Recently, People Accepted the 'Fact' of Aliens in the Solar System." *Scientific American,* 2 Feb. 2012, https://www.scientificamerican.com/article/until

-recently-people-accepted-the-fact-of-aliens-in-the-solar-system
Scharf, Caleb, and Leroy Cronin. "Quantifying the origins of life on a planetary scale." *PNAS* 113, no. 29 (2016): 8127–8132.
Scharf, Caleb et al. "A Strategy for Origins of Life Research." *Astrobiology* 15, no. 12 (2015): 1031–1042.
Stableford, Brian. "Science fiction before the genre." *The Cambridge Companion to Science Fiction,* edited by Edward James. Cambridge University Press, 2003, 15–31.
Traphagan, John W. *Extraterrestrial Intelligence and Human Imagination*. Springer, 2014.
Van Kranendonk, Martin, David Deamer, and Tara Djokic. "Life Springs." *Scientific American,* July 2017.
Walker, Sara Imari. "Origins of life: a problem for physics, a key issues review." *Reports on Progress in Physics* 80, no. 9 (2017).
Walker, Sara Imari. "We Need to Change How We Search for Alien Life." *Slate,* 23 Dec. 2020, https://slate.com/technology/2020/12/venus-phosphine-astrobiology-crisis-alien-life.html
Walker, Sara Imari, and Paul C.W. Davies. "The algorithmic origins of life." *Journal of the Royal Society Interface* 10 (2012).

Patruno, A., and M. Kama. "Neutron star planets: Atmospheric processes and irradiation." *Astronomy & Astrophysics* 608 (2017).

Sagan, Carl et al. "A search for life on Earth from the Galileo spacecraft." *Nature* 365 (1993): 715–721.

Wall, Mike. "Kepler-452b: What It Would Be Like to Live On Earth's 'Cousin.'" Space.com, 24 July 2016, https://www.space.com/30034-earth-cousin-exoplanet-kepler-452b-life.html

Chapter 3: Animals

"Anomalocaris." *Shape of Life,* https://www.shapeoflife.org/news/featured-creature/2018/02/26/anomalocaris

"Patrick Matthew." *Early Evolutionists,* https://early-evolution.oeb.harvard.edu/patrick-matthew

"The Enemy Within." *Star Trek,* 1x04, 6 Oct. 1966.

Avatar. James Cameron, director. 20th Century Fox, 2009.

Battaglia, Andy. "The Man Who Draws Dinosaurs." *The New Yorker,* 5 Dec. 2013, http://www.newyorker.com/culture/culture-desk/the-man-who-draws-dinosaurs

Burke, Sue. *Semiosis.* Tor Books, 2018.

Cohen, Jeffrey Jerome. "Monster Culture

(Seven Theses)." *Monster Theory: Reading Culture,* edited by Jeffrey Jerome Cohen. University of Minnesota Press, 1996.

Conway Morris, Simon. *Life's Solution: Inevitable Humans in a Lonely Universe*. Cambridge University Press, 2003.

Crichton, Michael. *Sphere*. Vintage, 2012.

Gould, Stephen Jay. *Wonderful Life: The Burgess Shale and the Nature of History*. W. W. Norton & Company, 1990.

Losos, Jonathan. *Improbable Destinies: Fate, Chance, and the Future of Evolution*. Riverhead, 2017.

Pullman, Philip. *The Amber Spyglass*. Alfred A. Knopf Books for Young Readers, 2007.

Sagan, Carl. *Cosmos*. Ballantine Books, 2013.

Sherman, Paul W. et al. "Naked Mole Rats." *Scientific American* 267, no. 2 (1992): 72–79.

Sherman, Paul W., Jennifer U.M. Jarvis, and Richard D. Alexander, eds. *The Biology of the Naked Mole-Rat*. Princeton University Press, 2017.

Shermer, Michael. "The Meaning of Life in a Formula." *Scientific American* 313, no. 2 (2015): 83.

Slivensky, Katie. "Alien Animals." *Discoverific!,* http://discoverific.blogspot.com/2012/04/alien-animals.html

Chapter 4: People

Abumrad, Jad, and Robert Krulwich, hosts. "Colors." *Radiolab,* WNYC, 21 May 2012, https://radiolab.org/episodes/211119-colors

Arrival. Denis Villeneuve, director. Paramount Pictures Studios, 2016.

Burke, Sue. *Semiosis*. Tor Books, 2018.

Butler, Octavia E. *Dawn*. Aspect, 1997.

Cambias, James L. *A Darkling Sea*. Macmillan, 2014.

Chiang, Ted. "Story of Your Life." *Stories of Your Life and Others*. Knopf, 2010.

Conway Morris, Simon. *Life's Solution: Inevitable Humans in a Lonely Universe*. Cambridge University Press, 2003.

Eiseley, Loren. "The Long Loneliness: Man and the Porpoise: Two Solitary Destinies." *The American Scholar* 30, no. 1 (1960): 57–64

Foster, Charles. *Being a Beast: Adventures Across the Species Divide*. Macmillan, 2016.

Grebowicz, Margret. *Whale Song*. Bloomsbury Academic, 2020.

Hunter College. "Hunter@Home — The Dolphin in the Mirror: Reflections on Dolphin Intelligence & Communication." YouTube, 7 May 2020, https://www.youtube.com/watch?v=ETLhG3wZfoE

Hurst, Nathan. "How Does Human

Echolocation Work?" *Smithsonian Magazine,* 2 Oct. 2017, https://www.smithsonianmag.com/innovation/how-does-human-echolocation-work-180965063/

Johnson, George. "The Battle for the Great Apes." *Pacific Standard,* 21 Nov. 2016, https://psmag.com/news/the-battle-for-the-great-apes-inside-the-fight-for-non-human-rights

Keim, Brandon. "An Elephant's Personhood on Trial." *The Atlantic,* 28 Dec. 2018, https://www.theatlantic.com/science/archive/2018/12/happy-elephant-personhood/578818/

Langlois, Krista. "When Whales and Humans Talk." *Hakai Magazine,* 3 Apr. 2018, https://hakaimagazine.com/features/when-whales-and-humans-talk/

Le Guin, Ursula K. "The Author of the Acacia Seeds." *The Unreal and the Real*. Saga Press, 2017.

Lem, Stanisław. *Solaris*. Houghton Mifflin Harcourt, 2002.

Marino, Lori. "The landscape of intelligence." *The Impact of Discovering Life Beyond Earth,* edited by Steven J. Dick. Cambridge University Press, 2015.

Nagel, Thomas. "What Is It Like to Be a Bat?" *The Philosophical Review* 83, no. 4 (1974): 435–450.

Oberhaus, Daniel. *Extraterrestrial Languages*. MIT Press, 2019.

Rowlands, Mark. "Are animals persons?" *Animal Sentience* 10, no. 1 (2016).

Sassi, Maria Michela. "The sea was never blue." *Aeon,* 31 July 2017, https://aeon.co/essays/can-we-hope-to-understand-how-the-greeks-saw-their-world

Chapter 5: Technology

"'Oumuamua." NASA.gov, https://solarsystem.nasa.gov/asteroids-comets-and-meteors/comets/oumuamua/in-depth/

"Relics." *Star Trek: The Next Generation,* 6x04, 12 Oct. 1992.

"The Second Renaissance." Mahiro Maeda, director. *The Animatrix*. Warner Bros. Home Entertainment, 2003.

Butler, Samuel. "Darwin Among the Machines." 13 June 1863, https://nzetc.victoria.ac.nz/tm/scholarly/tei-ButFir-t1-g1-t1-g1-t4-body.html

Chiang, Ted. "Why Computers Won't Make Themselves Smarter." *The New Yorker,* 30 Mar. 2021, https://www.newyorker.com/culture/annals-of-inquiry/why-computers-wont-make-themselves-smarter

Ćirković, M.M. "Kardashev's classification at 50+: A fine vehicle with room for improvement." *Serbian Astronomical Journal*

191 (2015): 1–15.
Cuthbertson, Anthony. “Elon Musk claims AI will overtake humans ‘in less than five years.’” *The Independent,* 27 July 2020, https://www.independent.co.uk/tech/elon-musk-artificial-intelligence-ai-singularity-a9640196.html
Denning, Kathryn. “‘L’ on Earth.” 56th International Astronautical Congress of the International Astronautical Federation, the International Academy of Astronautics, and the International Institute of Space Law, Oct. 2005, Fukuoka, Japan.
Denning, Kathryn. “Social Evolution.” *Cosmos and Culture,* edited by Steven J. Dick and Mark L. Lupisella. Government Printing Office, 2012.
Denning, Kathryn. “Ten thousand revolutions: conjectures about civilizations.” *Acta Astronautica* 68, no. 3–4 (2011): 381–388.
Denning, Kathryn, and Anamaria Berea. “Figuring Out, and Figuring In, the Human: Insights for Astrobiology from the Human Sciences.” For session: “Social Sciences, Philosophy, and History for Astrobiological Science.” AbSciCon 2019, 28 June 2019, Bellevue, Washington, USA.
Dick, Steven J. “Bringing Culture to Cosmos.” *Cosmos and Culture,* edited by Steven J. Dick and Mark L. Lupisella

Government Printing Office, 2012.
Dick, Steven J. "Cosmic Evolution." *Cosmos and Culture,* edited by Steven J. Dick and Mark L. Lupisella. Government Printing Office, 2012.
Frank, Adam. *Light of the Stars: Alien Worlds and the Fate of the Earth.* W. W. Norton & Company, 2018.
Frank, Adam et al. "The Anthropocene Generalized: Evolution of Exo-Civilizations and Their Planetary Feedback." *Astrobiology* 18, no. 5 (2018): 503–518.
Gardner, James. "The Intelligent Universe." *Cosmos and Culture,* edited by Steven J. Dick and Mark L. Lupisella. Government Printing Office, 2012.
Le Guin, Ursula K. "A Man of the People." *The Found and the Lost.* Simon and Schuster, 2016.
Mann, Adam. "Intelligent Ways to Search for Extraterrestrials." *The New Yorker,* 3 Oct. 2019, https://www.newyorker.com/science/elements/intelligent-ways-to-search-for-extraterrestrials
Sagan, Carl. "The Quest for Extraterrestrial Intelligence." *Cosmic Search* 1, no. 2 (1979), http://www.bigear.org/CSMO/HTML/CS02/cs02p02.htm
Stapledon, Olaf. *Star Maker.* Wesleyan University Press, 2004.

Swift, David W. *SETI Pioneers: Scientists Talk about Their Search for Extraterrestrial Intelligence*. University of Arizona Press, 1993.

Traphagan, John W. "Equating culture, civilization, and moral development in imagining extraterrestrial intelligence: anthropocentric assumptions?" *The Impact of Discovering Life beyond Earth,* edited by Steven J. Dick. Cambridge University Press, 2015.

Vinge, Vernor. "The coming technological singularity." *Whole Earth Review* (1993), https://frc.ri.cmu.edu/~hpm/book98/com.ch1/vinge.singularity.html

Vinge, Vernor. *A Fire Upon the Deep*. Macmillan, 1993.

Wright, Jason. "Dyson Spheres." *Serbian Astronomical Journal* 200 (2020): 1–18.

Wright, Jason. "The Ĝ Search for Kardashev Civilizations." *AstroWright,* https://sites.psu.edu/astrowright/the-g-hat-search-for-kardashev-civilizations/

Wright, Jason T., and Michael P. Oman-Reagan. "Visions of Human Futures in Space and SETI." *International Journal of Astrobiology* 17, no. 2 (2018): 177–188.

Chapter 6: Contact

"Carl Sagan discusses the book 'Contact.'" Studs Terkel Radio Archive

WFMT, originally broadcast 4 Oct. 1985, https://studsterkel.wfmt.com/programs/carl-sagan-discusses-book-contact
"Encyclopaedia Galactica." *Cosmos* part 12, 14 Dec. 1980.
"Insight from Afar." *National Jesuit News,* June 1998, https://marydoriarussell.net/novels/the-sparrow/national-jesuit-news-interview/
"It's the 25th anniversary of Earth's first (and only) attempt to phone E.T." *Cornell News,* 12 Nov. 1999, https://web.archive.org/web/20080802005337/http://www.news.cornell.edu/releases/Nov99/Arecibo.message.ws.html
"Reducing the Likelihood of Future Human Activities That Could Affect Geologic High-Level Waste Repositories." Office of Nuclear Waste Isolation, 1984.
"What are the contents of the Golden Record?" NASA, https://voyager.jpl.nasa.gov/golden-record/whats-on-the-record/
Abumrad, Jad, and Robert Krulwich, hosts. "Space." *Radiolab,* WNYC, 19 Aug. 2010, https://radiolab.org/episodes/91850-ann-druyen-on-the-space-episode
Beauchamp, Scott. "How to Send a Message 1,000 Years to the Future." *The Atlantic,* 24 Feb. 2015, https://www.theatlantic.com/technology/archive/2015/02

/how-to-send-a-message-1000-years-to-the-future/385720/

Billings, Linda. "Astrobiology in Culture: The Search for Extraterrestrial Life as 'Science.'" *Astrobiology* 12, no. 10 (2012).

Billings, Linda. "From Earth to the Universe: Life, Intelligence, and Evolution." *Biological Theory* 13 (2018): 93–102.

Burke, Sue. "Let ME talk to the aliens! Ted Chiang's 'Story of Your Life.'" *Tor.com,* 6 Feb 2018, https://www.tor.com/2018/02/06/let-me-talk-to-the-aliens-ted-chiangs-story-of-your-life/

Calvey, Ryan. *Transcendent Outsiders, Alien Gods, and Aspiring Humans: Literary Fantasy and Science Fiction as Contemporary Theological Speculation*. The Graduate School, Stony Brook University, PhD dissertation, 2011.

Chiang, Ted. "Story of Your Life." *Stories of Your Life and Others*. Knopf, 2010.

Clinton, Bill. "President Clinton Statment Regarding Mars Meteorite Discovery." Office of the Press Secretary, The White House, 7 Aug. 1996, https://www2.jpl.nasa.gov/snc/clinton.html

Cocconi, G., and P. Morrison. "Searching for Interstellar Communications." *Nature* 184 (1959): 844–846.

Contact. Robert Zemeckis, director. Warner

Bros. Pictures, 1997.
Cooper, Keith. *The Contact Paradox: Challenging Our Assumptions in the Search for Extraterrestrial Intelligence*. Bloomsbury Sigma, 2019.
Danis, Sam. "Author Tade Thompson On The 'Frankenstein of Influences' That Helped Create His Buzzy Sci-fi Debut 'Rosewater.'" *Audible Blog,* 9 Oct. 2018, https://www.audible.com/blog/author-tade-thompson-rosewater-frankenstein-of-influences
Denning, Kathryn. "Being technological." *Acta Astronautica* 68, no. 3–4 (2011): 372–380.
Denning, Kathryn. "Ten thousand revolutions: conjectures about civilizations." *Acta Astronautica* 68, no. 3–4 (2011): 381–388.
Denning, Kathryn et al. "SETI and Post-Detection: Towards a New Research Roadmap." 70th International Astronautical Congress, 21–25 Oct. 2019.
Dick, Steven J. "History, discovery, analogy." *The Impact of Discovering Life beyond Earth,* edited by Steven J. Dick. Cambridge University Press, 2015.
FitzPatrick, Jessica. "Seeds of Catastrophe: 'The Rosewater Insurrection' and 'The Rosewater Redemption.'" *Los Angeles Review of Books,* 29 Feb. 2020, https://

lareviewofbooks.org/article/seeds-of-catastrophe-the-rosewater-insurrection-and-the-rosewater-redemption/

FitzPatrick, Jessica. "Twenty-First Century Afrofuturist Aliens: Shifting to the Space of Third Contact." *Extrapolation* 61, no. 1 (2020).

Garber, Stephen J. "Searching for Good Science." *Journal of The British Interplanetary Society* 52 (1999): 3–12, https://history.nasa.gov/garber.pdf

Gevers, Nick. "Of Prayers and Predators." *Infinity Plus,* 1999, http://www.infinityplus.co.uk/nonfiction/intmdr.htm

Gliedman, John. "Things No Amount of Learning Can Teach." *Omni* 6, no. 11 (1983), https://chomsky.info/198311__/

Greaves, J.S. et al. "Phosphine gas in the cloud decks of Venus." *Nature Astronomy* 5 (2021): 655–664.

Hauser, Marc D., Noam Chomsky, and W. Tecumseh Fitch. "The Faculty of Language: What Is It, Who Has It, and How Did It Evolve?" *Science* 298, no. 5598 (2002): 1569–1579.

Haywood Ferreira, Rachel. "Second Contact: The First Contact Story in Latin American Science Fiction." *Parabolas of Science Fiction,* edited by Brian Attebery and Veronica Hollinger. Wesleya

University Press, 2013, 70–88.
Howard, Kat. "Interview: Mary Doria Russell." *Lightspeed* 14 (2011), https://www.lightspeedmagazine.com/nonfiction/feature-interview-mary-doria-russell/
Lempert, William. "From Interstellar Imperialism to Celestial Wayfinding: Prime Directives and Colonial Time-Knots in SETI." *American Indian Culture and Research Journal* 45, no. 1 (2021): 45–70.
Marantz, Alec. "What do linguists do?" Feb. 2019.
Miéville, China. *Embassytown*. Random House Digital, Inc., 2012.
Morena, Anthony Michael. *The Voyager Record: A Transmission*. Rose Metal Press, 2016.
Moro, Andrea. *Impossible Languages*. MIT Press, 2016.
Murphy, Ricky Leon. "The ALH 84001 Controversy." AstronomyOnline.org, http://astronomyonline.org/Astrobiology/ALH84001.asp#R3
Neimark, Jill. "God, Baseball, and Science: An Interview With Mary Doria Russell." *Metanexus,* 19 Jan. 2003, https://metanexus.net/god-baseball-and-science-interview-mary-doria-russell/
Okorafor, Nnedi. *Lagoon*. Simon and Schuster, 2016.

Piesing, Mark. "How to build a nuclear warning for 10,000 years' time." 3 Aug. 2020, https://www.bbc.com/future/article/20200731-how-to-build-a-nuclear-warning-for-10000-years-time

Russell, Mary Doria. *The Sparrow.* Ballantine Books, 2008.

Sagan, Carl. *Contact.* Gallery Books, 2019.

Samatar, Sofia. "An Interview with Tade Thompson." *Interfictions Online* 7 (2016), http://interfictions.com/an-interview-with-tade-thompson/

Sample, Ian. "Scientists looking for aliens investigate radio beam 'from nearby star.'" *The Guardian,* 18 Dec. 2020, https://www.theguardian.com/science/2020/dec/18/scientists-looking-for-aliens-investigate-radio-beam-from-nearby-star

Schulze-Makuch, Dirk. "All Eyes on Alpha Centauri." *Air & Space Magazine,* 14 Apr. 2012, https://www.airspacemag.com/daily-planet/all-eyes-alpha-centauri-180977507/

Sebeok, Thomas A. "Communication Measures to Bridge Ten Millennia." U.S. Department of Energy Office of Scientific and Technical Information, 1 Apr. 1984.

Skrewtape. "Mark Okrand on Klingon." YouTube, 3 May 2012, https://www.youtube.com/watch?v=e5Did-eVQDc

Swift, David W. *SETI Pioneers: Scientists Talk about Their Search for Extraterrestrial Intelligence.* University of Arizona Press, 1993.

Talks at Google. "Living Language Dothraki | David Peterson | Talks at Google." YouTube, 18 Nov. 2014, https://www.youtube.com/watch?v=SUUHH5lpLL4

Tavares, Frank. "Ethical Exploration and the Role of Planetary Protection in Disrupting Colonial Practices." A submission to the Planetary Science and Astrobiology Decadal Survey 2023–2032, https://drive.google.com/file/d/1ca8RRy1MSpOAvexucgxWJIlBuNsdPRn8/view

The Staff at the National Astronomy and Ionosphere Center. "The Arecibo message of November, 1974." *Icarus* 26, no. 4 (1975): 462–466.

Thompson, Tade. *Rosewater.* Orbit, 2017.

Thornton, Jonathan. "Interview with Tade Thompson." *The Fantasy Hive,* 26 Nov. 2018, https://fantasy-hive.co.uk/2018/11/interview-with-tade-thompson/

Trauth, Kathleen M., Stephen C. Hora, and Robert V. Guzowski. "Expert Judgment on Markers to Deter Inadvertent Human Intrusion into the Waste Isolation Pilot Plant." Sandia National Laboratories, 1993.

Walkowicz, Lucianne. "I'm a Little Worried About Venus." *Slate,* 15 Sept. 2020, https://slate.com/technology/2020/09/venus-phosphine-life-planetary-protection.html

Hopeful Monsters

Booth, Austin, and W. Ford Doolittle. "Eukaryogenesis, how special really?" *PNAS* 112, no. 33 (2015): 10278–10285.

Callier, Viviane. "DNA's Histone Spools Hint at How Complex Cells Evolved." *Quanta Magazine,* 10 May 2021, https://www.quantamagazine.org/dnas-histone-spools-hint-at-how-complex-cells-evolved-20210510

Kasting, James F. "Peter Ward and Donald Brownlee's Rare Earth." *Perspectives in Biology and Medicine* 44, no. 1 (2001): 117–131.

Lane, Nick. *The Vital Question: Energy, Evolution, and the Origins of Complex Life*. W. W. Norton & Company, 2015.

Lane, Nick, and William F. Martin. "Eukaryotes really are special, and mitochondria are why." *PNAS* 112, no. 35 (2015): E4823.

L'Engle, Madeleine. *A Wind in the Door*. Macmillan, 1973.

L'Engle, Madeleine. *A Wrinkle in Time*. Farrar

Straus and Giroux (BYR), 2010.
Ward, Peter D., and Donald Brownlee. *Rare Earth: Why Complex Life is Uncommon in the Universe*. Copernicus, 2003.

ABOUT THE AUTHOR

Jaime Green is a science writer, essayist, editor, and teacher, and she is series editor of *The Best American Science and Nature Writing*. She received her MFA in Creative Nonfiction from Columbia, and her writing has appeared in *Slate, Popular Science, The New York Times Book Review, American Theatre, Catapult, Astrobites,* and elsewhere. She lives in Connecticut with her husband and son.

The employees of Thorndike Press hope you have enjoyed this Large Print book. All our Thorndike Large Print titles are designed for easy reading, and all our books are made to last. Other Thorndike Press Large Print books are available at your library, through selected bookstores, or directly from us.

For information about titles, please call:

(800) 223-1244

or visit our website at:

http://gale.cengage.com/thorndike